Mathematical Description of Linear Systems

Volume 1: Discrete Techniques of Parameter Estimation: The Equation Error Formulation, JERRY M. MENDEL

Volume 2: Mathematical Description of Linear Systems, WILSON J. RUGH

OTHER VOLUMES IN PREPARATION

Mathematical Description of Linear Systems

Wilson J. Rugh

Department of Electrical Engineering
The Johns Hopkins University
Baltimore, Maryland

MARCEL DEKKER, INC. New York and Basel

To

Mr. and Mrs. DeWitt W. Rugh
Mr. and Mrs. Hervey N. Saul

my grandparents

MARCEL DEKKER, INC.

270 Madison Avenue, New York, New York 10016

LIBRARY OF CONGRESS CATALOG CARD NUMBER: 75-1684

ISBN: 0-8247-6300-9

Current printing (last digit)
10 9 8 7 6 5 4 3 2

PRINTED IN THE UNITED STATES OF AMERICA

PREFACE

This book is intended to provide a brief, highly organized introductory treatment of the mathematical description of constant parameter linear systems. The emphasis is on a consistent point of view which attaches equal importance to the internal description and the input/output description and which delineates their respective roles in the theory of linear systems. An algebraic approach which reflects current trends in the research literature is used.

The material in the book is self-contained and presented in a reasonably rigorous manner. Almost all results are proved. For those which are not, the Remarks and References section in each chapter lists appropriate sources. The level of presentation is suitable for upper division undergraduates and beginning graduate students in engineering and the quantitative sciences. In addition to a standard undergraduate calculus sequence, the major prerequisite for the material is a working knowledge of linear algebra and matrix operations. Some exposure to signal, circuit, or system ideas is helpful, although mathematical maturity can be substituted for this engineering background.

Although the Table of Contents serves as a topical outline, several specific comments about the material should be made. Continuous- and discrete-time systems are treated side-by-side in the first few chapters. This parallel treatment is gradually loosened until, in the later chapters, only one case is discussed and the other is left to the Problems section.

Both the z- and s-transforms are introduced as algebraic entities using the theory of Hankel matrices and rational functions. Although this formulation has both advantages and disadvantages in comparison with the more traditional approach, it does fit especially well with the modern treatment of system realization and identification.

Contrary to most treatments of linear system theory, the unit impulse (generalized) function is not discussed. To do so in a rigorous fashion would require a digression not warranted at the level of this book. The omission of the topic costs some mathematical convenience but little operational significance or usefulness. We do consider two topics which lead to the impulse function: the response of a linear system to unit area pulses, and a description of sampling.

Much of the material on complete identification is not standard. The treatment is intended to expose the basic structural features of this very important topic. We also point out aspects of the identification problem which are of current research interest.

A small collection of simple examples is introduced in the first chapter and used repeatedly to illustrate topics and provide problems. Of course additional examples culled from texts or the reader's educational background can facilitate understanding. Classroom discussion of examples relevant to the interests of the class can enhance considerably the mathematical concepts and techniques of system description.

A natural follow-up to the material in this book would be a study of analysis, design, and optimization techniques in linear systems. Methods based upon the state vector equation as well as the transfer function description could be discussed and compared. Such material would provide both review and applications of the topics discussed herein.

This book was developed from the first semester of a course entitled *Systems*, which I have taught for a number of years at The Johns Hopkins University. The enrollment has consisted of juniors and seniors in electrical engineering and a number of other disciplines along with some beginning graduate students. To this perceptive audience I owe many thanks. For stimulating discussion and criticism, I thank

Professors W. H. Huggins and G. G. L. Meyer of Johns Hopkins and Professor E. L. Titlebaum of the University of Rochester. Also the diligent efforts of Mrs. Gloria Baker and Mrs. Margery Neikirk in preparing the numerous iterations of the manuscript are gratefully acknowledged.

Wilson J. Rugh
Baltimore, Maryland

ABBREVIATIONS AND NOTATION

(A,b,c,d)	coefficients of a state vector equation
CR	completely reachable
CO	completely observable
RCF	reachability canonical form
OCF	observability canonical form
AS	asymptotically stable
UBIBOS	uniformly bounded input bounded output stable
i	$\sqrt{-1}$
$\binom{n}{k}$	binomial coefficient
iff	if and only if
det	determinant
$(\ )'$	transpose of ( )
$(\dot{\ })$	$\frac{d}{dt}(\ )$
$(\ )^{(i)}$	$\frac{d^i}{dt^i}(\ )$
$(\bar{\ })$	complex conjugate of ( )
H_f	Hankel matrix corresponding to f
$u_0(k)$	discrete-time unit pulse
$u_{-1}(k)$	discrete-time unit step
$u_{-2}(k)$	discrete-time unit ramp
$u_{-1}(t)$	continuous-time unit step
$u_{-2}(t)$	continuous-time unit ramp
Z[ ]	z-transform of [ ]
S[ ]	s-transform of [ ]
Re( )	real part of ( )
Im( )	imaginary part of ( )

$[a_1 | a_2 | \ldots | a_n]$ column partitioned matrix

$$\begin{bmatrix} b_1 \\ \hline b_2 \\ \hline \vdots \\ \hline b_n \end{bmatrix}$$ row partitioned matrix

$$\begin{bmatrix} a_1 & & & \\ & a_2 & & \\ & & \ddots & \\ & & & a_n \end{bmatrix}$$ diagonal matrix

CONTENTS

CHAPTER 1

INTERNAL SYSTEM DESCRIPTION: THE STATE VECTOR EQUATION

In this chapter we introduce the basic mathematical description of the internal structure of linear systems. This description requires that we know precisely how all parts of the system behave, how they are interconnected, and how they interact. Of course this detailed information is usually provided at the expense of severe assumptions on the behavior of the system at hand.

For both continuous-time and discrete-time systems, we always arrange the mathematical description of internal structure into a standard form of a vector first-order differential or difference equation. This is the state vector equation. We can use different variables to describe a given system so the effect of certain types of variable changes is considered. A particular variable change which yields a canonical form for the state vector equation is also discussed.

We conclude the chapter with a treatment of the basic solution of the state vector equation and the properties of this solution in both the continuous and discrete-time cases.

1.1. THE DISCRETE-TIME STATE VECTOR EQUATION

We regard discrete-time signals as sequences of vectors or scalars defined for integers k with the convention that for all $k < 0$ the signals have the value 0. This so-called *one-sided signal convention* is commonly used since most processes have a fixed beginning which we denote by $k = 0$. The notation $f(k)$ will be used to denote either the k-th value of the signal or the entire signal. The precise meaning will be clear from the context. By addition and scalar multiplication of discrete-time signals we mean the usual element-wise addition and scalar multiplication of sequences.

A collection of discrete-time signals will be described as a vector composed of the signals, that is, a vector signal. Addition of two vector signals (of the same dimension) is defined in the usual component-wise fashion. Multiplication of a vector signal by a constant matrix (of appropriate dimension) also is defined in the usual way. The result of such a multiplication is a new vector signal whose i-th component signal is a linear combination of the components of the original vector signal.

We will consider linear discrete-time state vector equations of the form

$$\begin{aligned} x(k+1) &= Ax(k) + bu(k), \quad x(0) = x_o \\ y(k) &= cx(k) + du(k), \quad k = 0, 1, 2, \ldots \end{aligned} \tag{1.1}$$

The signal $u(k)$ is the input and $y(k)$ is the output. Both are assumed to be scalar signals. The $n \times 1$ vector signal $x(k)$ is the state vector composed of the n state variables $x_1(k)$, $x_2(k)$, ..., $x_n(k)$, and x_o is the initial state. Because of the one-sided signal convention, we always take $k \geq 0$ without loss of generality.

The coefficient matrix A is $n \times n$, b is $n \times 1$, c is $1 \times n$, and d is a scalar. These are all composed of real numbers unless otherwise noted. We will have occasion to use complex coefficients in the sequel. Often we denote the state vector equation (1.1) by the 4-tuple of coefficient matrices (A,b,c,d).

A state vector equation description can arise in many different situations as the following examples indicate. One of the important features of these examples to note is the role played by the assumptions on system behavior.

Example 1 Natchez Indian Social Structure. The Natchez Indian culture is divided into four classes: Suns, Nobles, Honoreds, and Stinkards. A complete table of the allowable marriages and the class of the resulting offspring is given in Table 1-1.

TABLE 1-1
Natchez Marriage Rules

Mother	Father	Offspring
Sun	Stinkard	Sun
Noble	Stinkard	Noble
Honored	Stinkard	Honored
Stinkard	Sun	Noble
Stinkard	Noble	Honored
Stinkard	Honored	Stinkard
Stinkard	Stinkard	Stinkard

The growth of the population of each particular class is of interest to anthropologists. To obtain a simple model, we make the following assumptions.

1. Each class has an equal number of men and women
2. Each individual marries once and only once
3. Each couple has one son and one daughter

These assumptions imply that the total population remains constant in each generation and that the female population distribution is identical to the male population distribution. Thus we need only consider the male population distribution.

Let $x_1(k)$ be the number of male Suns in generation k, $x_2(k)$ the number of male Nobles in generation k, $x_3(k)$ the number of male Honoreds in generation k, and $x_4(k)$ the number of male Stinkards in generation k. Since a Sun son is produced by every Sun mother and in no other way and since the number of Sun mothers is the same as the number of Sun males, we can write

$$x_1(k+1) = x_1(k)$$

A Noble son is produced by every Sun father and every Noble mother, but by no other parents. Thus we obtain

$$x_2(k+1) = x_1(k) + x_2(k)$$

An Honored son is produced by every Honored mother and by every Noble father, and in no other way. Thus

$$x_3(k+1) = x_2(k) + x_3(k)$$

To obtain an equation for the Stinkard population, note that the total population remains constant so that

$$\begin{aligned} x_4(k+1) &= x_1(k) + x_2(k) + x_3(k) + x_4(k) - x_1(k+1) - x_2(k+1) - x_3(k+1) \\ &= -x_1(k) - x_2(k) + x_4(k) \end{aligned}$$

We can arrange these equations into the form of a state vector equation to obtain

$$\begin{bmatrix} x_1(k+1) \\ x_2(k+1) \\ x_3(k+1) \\ x_4(k+1) \end{bmatrix} = \begin{bmatrix} 1 & 0 & 0 & 0 \\ 1 & 1 & 0 & 0 \\ 0 & 1 & 1 & 0 \\ -1 & -1 & 0 & 1 \end{bmatrix} \begin{bmatrix} x_1(k) \\ x_2(k) \\ x_3(k) \\ x_4(k) \end{bmatrix} \qquad k = 0,1,2,\ldots \tag{1.2}$$

Of course a population distribution must be specified in the initial generation, $k = 0$.

Example 2 A Fish Hatchery. A simple mathematical model of a fish hatchery can be derived as follows. We assume there are four stages of growth of the fish: eggs, fry, young, and adults. The input to the hatchery is a number of fish eggs each year, and the output is the number of young fish removed from the hatchery each year. Let $u(k)$ be the number of eggs supplied in year k, $x_1(k)$ the number of fry in year k, $x_2(k)$ the number of young in year k, and $x_3(k)$ the number of adults in year k.

The number of fry in year $k + 1$ is given by the number of eggs produced by adults in year k, less the number of eggs eaten by the fry and young in year k, plus the number of eggs supplied in year k. Assuming these quantities are expressible as fractions of the various populations, we can write

$$x_1(k+1) = a_1x_3(k) - a_2x_2(k) - a_3x_1(k) + u(k)$$

The number of young in year k + 1 is assumed to be equal to the number of fry in year k, less the number of young removed from the hatchery in year k,

$$x_2(k+1) = x_1(k) - a_4x_2(k)$$

The number of adults in year k + 1 is the number of adults in year k, plus the number of young remaining from year k, less the number of adults that die in year k. Again, we assume this can be described by a linear equation.

$$x_3(k+1) = a_5x_2(k) + (1 - a_6)x_3(k)$$

Letting x(k) be the 3 × 1 vector with components $x_1(k)$, $x_2(k)$, and $x_3(k)$ and letting the output y(k) be the number of young fish removed in year k, we can arrange these equations into a state vector equation of the form (1.1).

$$x(k+1) = \begin{bmatrix} -a_3 & -a_2 & a_1 \\ 1 & -a_4 & 0 \\ 0 & a_5 & 1-a_6 \end{bmatrix} x(k) + \begin{bmatrix} 1 \\ 0 \\ 0 \end{bmatrix} u(k) \qquad (1.3)$$

$$y(k) = \begin{bmatrix} 0 & a_4 & 0 \end{bmatrix} x(k)$$

According to the standard form we let x(0) be the initial stocking of the hatchery and k = 0, 1, 2, ...

Example 3 Difference Equations of n-th Order. The mathematical modeling of a system often results in an n-th order difference equation of the form

$$y(k+n) + a_{n-1}y(k+n-1) + \ldots + a_1y(k+1) + a_0y(k) = bu(k), \quad k = 0, 1, \ldots$$

with the initial conditions y(0), y(1),..., y(n-1) given. We shall regard this as an internal system description with input u(k), output y(k), and implicit internal variables y(k), y(k+1),..., y(k+n-1). This description can be written as a state vector equation by explicitly labeling these as

$$
\begin{aligned}
x_1(k) &= y(k) \\
x_2(k) &= y(k+1) \\
&\vdots \\
x_n(k) &= y(k+n-1)
\end{aligned}
\tag{1.4}
$$

In terms of these variables we can write

$$
\begin{aligned}
x_1(k+1) &= y(k+1) = x_2(k) \\
x_2(k+1) &= y(k+2) = x_3(k) \\
&\vdots \\
x_{n-1}(k+1) &= y(k+n-1) = x_n(k)
\end{aligned}
$$

and substituting the definitions in (1.4) into the n-th order difference equation,

$$x_n(k+1) = -a_0x_1(k) - a_1x_2(k) - \dots -a_{n-1}x_n(k) + bu(k)$$

Thus with $x(k)$ the column vector with components $x_1(k)$, $x_2(k)$, ..., $x_n(k)$, we have

$$
x(k+1) = \begin{bmatrix} 0 & 1 & 0 & \dots & 0 & 0 \\ 0 & 0 & 1 & \dots & 0 & 0 \\ \vdots & \vdots & \vdots & & \vdots & \vdots \\ 0 & 0 & 0 & \dots & 0 & 1 \\ -a_0 & -a_1 & -a_2 & \dots & -a_{n-2} & -a_{n-1} \end{bmatrix} x(k) + \begin{bmatrix} 0 \\ 0 \\ \vdots \\ 0 \\ b \end{bmatrix} u(k)
$$

$$
y(k) = [1 \quad 0 \quad 0 \quad \dots \quad 0]\, x(k), \quad k = 0, 1, \dots
\tag{1.5}
$$

From (1.4) the initial state vector is given by

$$
x(0) = \begin{bmatrix} y(0) \\ y(1) \\ \vdots \\ y(n-1) \end{bmatrix}
$$

1.2. DISCRETE-TIME STATE VARIABLE DIAGRAMS

A state variable diagram (SVD) is a pictorial representation of a state vector equation. An SVD is composed of three basic elements: adders, scalars, and unit delayors.

The output signal of an adder is the (signed) sum of the input signals. For example, see Fig. 1-1. At each k the

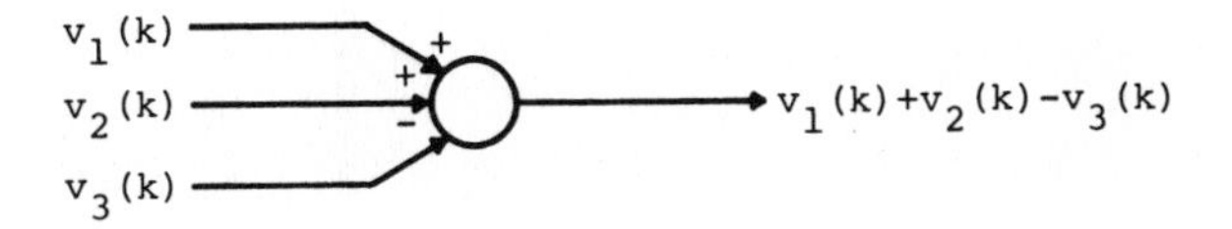

FIG. 1-1 An adder.

value of the output signal is $v_1(k) + v_2(k) - v_3(k)$.

The output signal of a scalar is simply the input signal multiplied by the indicated real number as shown in Fig. 1-2.

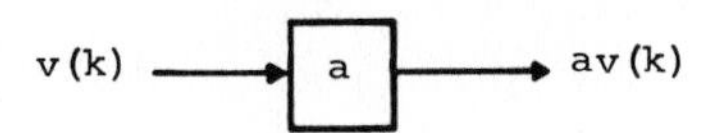

FIG. 1-2 A scalar.

The output signal of a unit delayor is the input signal "shifted to the right" by one unit. If the input signal is v(k), then the output signal w(k) is given by w(0) = v(-1) = 0, w(1) = v(0), w(2) = v(1), ... The one-sided signal convention implies that the output of a unit delayor always has the value 0 at k = 0. To remove this restriction we add an initial condition terminal so that any desired value of w(0) can be specified as in Fig. 1-3.

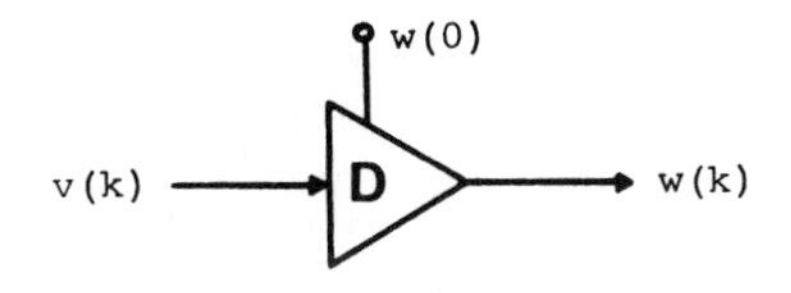

FIG. 1-3 A unit delayor.

The key property of the unit delayor is that it can be used to represent the relationship between the signals v(k) and v(k+1), that is, between the sequences

$$(v(0), v(1), v(2), \ldots)$$

and

$$(v(1), v(2), v(3), \ldots)$$

Clearly, if the second signal is shifted one unit to the right and v(0) is supplied, we obtain the first signal. In terms of the unit delayor, this relationship is indicated in Fig. 1-4.

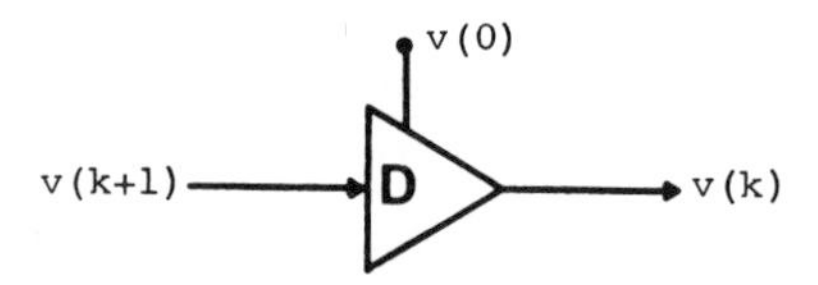

FIG. 1-4 A signal relationship involving the unit delayor.

We use these three elements to represent a state vector equation as follows. For each state variable $x_i(k)$, the relationship between $x_i(k+1)$ and $x_i(k)$ is represented as above using the unit delayor. Then these variables along with the input and output can be connected according to the state vector equation using scalars and adders. Of course a rearrangement is often necessary to obtain a neat SVD.

Note that if an SVD representation of a system is given, then it is straightforward to obtain a state vector equation description by labeling the unit delayor outputs as the state variables.

Example 4 The Natchez Indians. The SVD corresponding to (1.2) can be drawn as in Fig. 1-5.

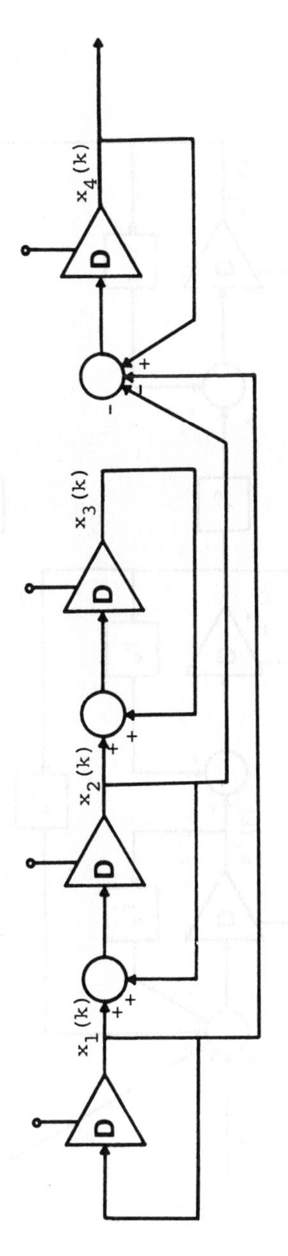

FIG. 1.5 The Natchez Indian SVD.

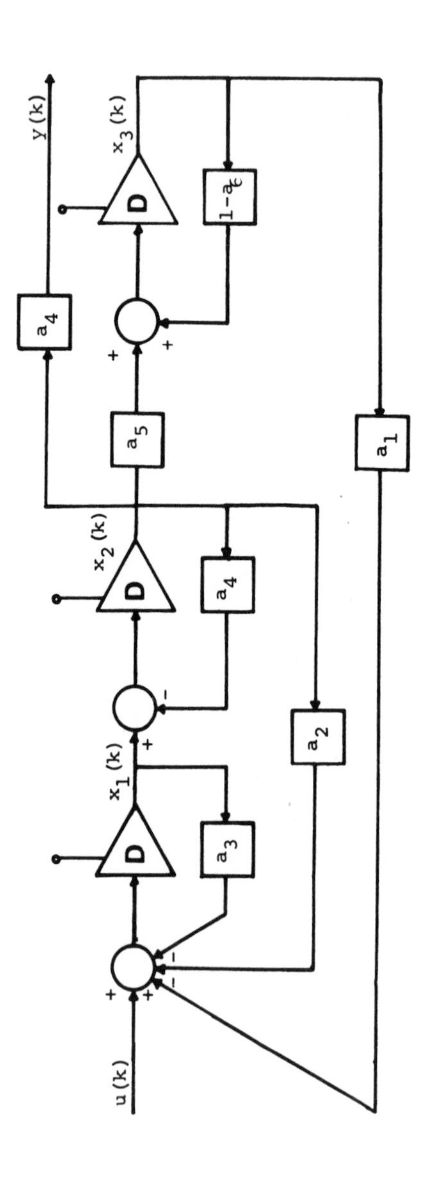

FIG. 1-6 The fish hatchery SVD.

Example 5 The Fish Hatchery. The SVD corresponding to (1.3) can be drawn as in Fig. 1-6.

1.3. THE CONTINUOUS-TIME STATE VECTOR EQUATION

A continuous-time signal is a function $f(t)$ defined for all real t with the one-sided signal convention that $f(t) = 0$ when $t < 0$. As in the discrete-time case, we use $f(t)$ to denote the entire signal or the value of the signal at t depending on the context. By addition and scalar multiplication, we mean the usual addition and scalar multiplication of functions.

Use will be made of the vector signal notation just as in the discrete-time case. Addition of vector signals and multiplication of a vector signal by a constant matrix are defined in the usual fashion.

We consider continuous-time state vector equations of the form

$$\begin{aligned} \dot{x}(t) &= Ax(t) + bu(t), \quad x(0) = x_o \\ y(t) &= cx(t) + du(t), \quad t \geq 0 \end{aligned} \tag{1.6}$$

The input $u(t)$ and the output $y(t)$ are scalar signals and $x(t)$ is the $n \times 1$ state vector signal with initial value x_o. The coefficient matrices A, b, c, and d are $n \times n$, $n \times 1$, $1 \times n$, and 1×1 with real elements unless specifically noted. Often (1.6) will be denoted by (A,b,c,d).

The formulation of a state vector equation for a continuous-time system is similar to the discrete-time case.

Example 6 Water Bucket System. We will often use water bucket systems constructed from the basic bucket shown in Fig. 1-7. The cross-sectional area of the bucket is C ft^2, and the depth of water is $x(t)$ ft. The basic assumption is that the flow out of the bucket is proportional to the water depth according to $y(t) = (1/R)x(t)$. The total volume of water in the bucket at any $t \geq 0$ is $Cx(t)$ ft^3, and the rate of change of volume is

$$C\dot{x}(t) = u(t) - y(t)$$

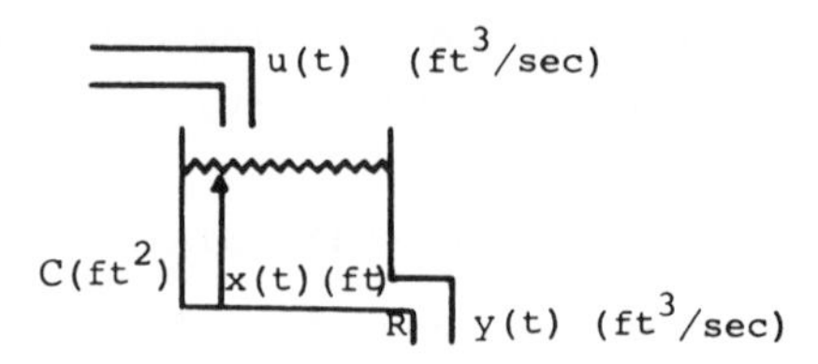

FIG. 1-7 A water bucket system.

Thus the single bucket system is described by the standard form equation

$$\dot{x}(t) = \frac{-1}{RC}\, x(t) + \frac{1}{C}\, u(t)$$

$$y(t) = \frac{1}{R}\, x(t)$$

Of course more complex water bucket systems are easily devised. For example, consider the parallel bucket system shown in Fig. 1-8. Assume that the units are the same as

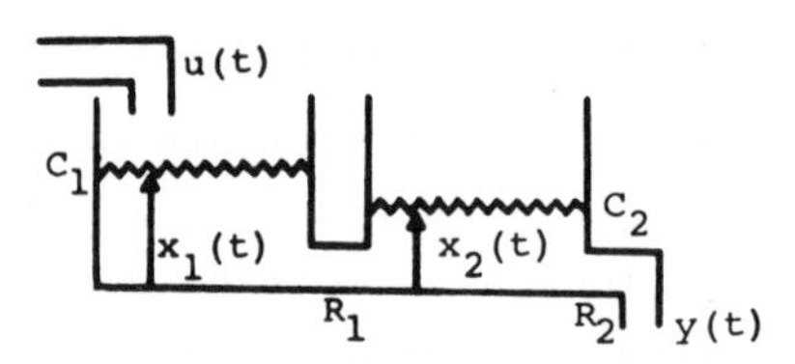

FIG. 1-8 A parallel water bucket system

in the single bucket system and that the flow through an orifice is proportional to the difference in water depth across the orifice. Then assuming the convention of positive flow from left to right, we have

$$C_1\dot{x}_1(t) = u(t) - \frac{1}{R_1}\, [x_1(t) - x_2(t)]$$

$$C_2\dot{x}_2(t) = \frac{1}{R_1}[x_1(t) - x_2(t)] - y(t)$$

$$y(t) = \frac{1}{R_2} x_2(t)$$

Letting x(t) be the 2 × 1 vector with components $x_1(t)$ and $x_2(t)$ yields the state vector equation

$$\dot{x}(t) = \begin{bmatrix} -1/(R_1C_1) & 1/(R_1C_1) \\ 1/(R_1C_2) & -1/(R_1C_2) - 1/(R_2C_2) \end{bmatrix} x(t) + \begin{bmatrix} 1/C_1 \\ 0 \end{bmatrix} u(t) \tag{1.8}$$

$$y(t) = \begin{bmatrix} 0 & 1/R_2 \end{bmatrix} x(t)$$

Example 7 RLC Electrical Circuits. To obtain a state vector equation for an electrical circuit, the following procedure can be used except for a few special situations. Label each inductor and capacitor with a voltage and a current and choose the inductor currents and capacitor voltages as the state variables. Then Kirchoff's Laws can be used to obtain equations of the appropriate form in terms of these variables.

Consider the circuit with voltage source input u(t) and output current $y(t) = i_2(t)$ shown in Fig. 1-9. From

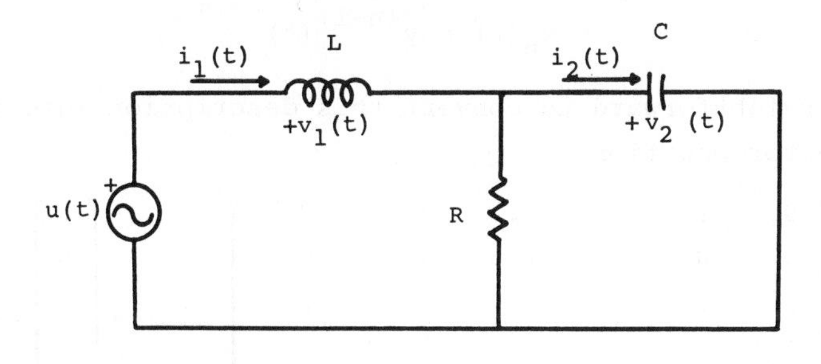

FIG. 1-9 An RLC circuit

Kirchoff's voltage and current laws we obtain, respectively,

$$L \frac{d}{dt} i_1(t) = u(t) - v_2(t)$$

$$C \frac{d}{dt} v_2(t) = i_1(t) - \frac{1}{R} v_2(t)$$

Letting $x(t)$ be the vector with components $i_1(t)$ and $v_2(t)$,

$$\dot{x}(t) = \begin{bmatrix} 0 & -1/L \\ 1/C & -1/(RC) \end{bmatrix} x(t) + \begin{bmatrix} 1/L \\ 0 \end{bmatrix} u(t) \qquad (1.9)$$

$$y(t) = [\,1 \quad -1/R\,]\, x(t)$$

Example 8 Differential Equations of n-th Order. Systems are often described in terms of a differential equation of the form

$$y^{(n)}(t) + a_{n-1}y^{(n-1)}(t) + \ldots + a_1 y^{(1)}(t) + a_0 y(t) = bu(t)$$

with initial conditions $y(0), y^{(1)}(0), \ldots, y^{(n-1)}(0)$. As in the discrete-time case, we view this as an internal description with the implicit internal variables $y(t), y^{(1)}(t), \ldots, y^{(n-1)}(t)$. With the definitions

$$\begin{aligned} x_1(t) &= y(t) \\ x_2(t) &= y^{(1)}(t) \\ &\vdots \\ x_n(t) &= y^{(n-1)}(t) \end{aligned} \qquad (1.10)$$

it is straightforward to convert this description into the state vector equation

$$\dot{x}(t) = \begin{bmatrix} 0 & 1 & 0 & \cdots & 0 & 0 \\ 0 & 0 & 1 & \cdots & 0 & 0 \\ \vdots & \vdots & \vdots & & \vdots & \vdots \\ 0 & 0 & 0 & \cdots & 0 & 1 \\ -a_0 & -a_1 & -a_2 & \cdots & -a_{n-2} & -a_{n-1} \end{bmatrix} x(t) + \begin{bmatrix} 0 \\ 0 \\ \vdots \\ 0 \\ b \end{bmatrix} u(t)$$

(1.11)

$$y(t) = \begin{bmatrix} 1 & 0 & \dots & 0 \end{bmatrix} x(t)$$

where

$$x_o = \begin{bmatrix} y(0) \\ y^{(1)}(0) \\ \vdots \\ y^{(n-1)}(0) \end{bmatrix}$$

1.4. CONTINUOUS-TIME STATE VARIABLE DIAGRAMS

The state variable diagram is also a useful representation for continuous-time state vector equations. In this case, an SVD is composed of adders, scalars, and integrators. The scalars and adders perform in the same manner as in the discrete-time case. Examples are shown in Fig. 1-10.

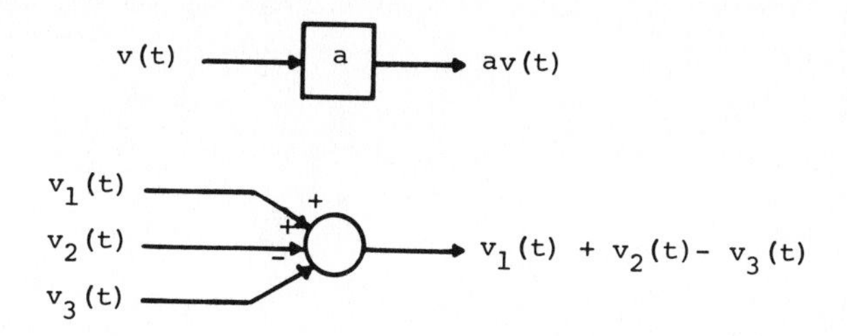

FIG. 1-10 Continuous-time scalar and adder.

An integrator yields an output signal which is the time integral of the input signal as shown in Fig. 1-11. Because

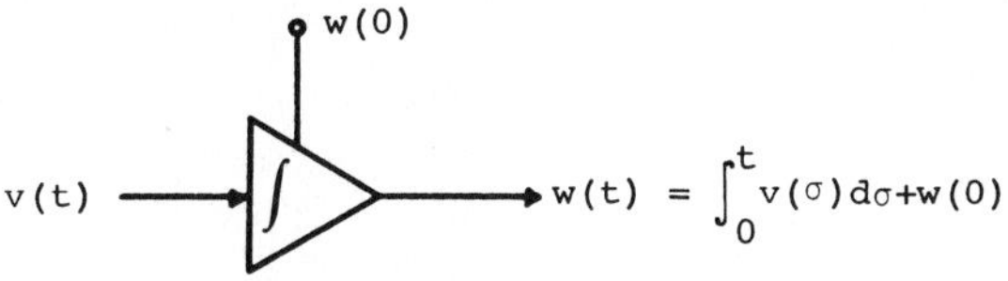

w(0)

v(t) $\int$ $w(t) = \int_0^t v(\sigma)\,d\sigma + w(0)$

FIG. 1-11 An integrator.

of the one-sided signal convention, we also include an initial condition terminal which sets the desired value of the output at t = 0. Using this device we can represent the relationship between a signal and its time derivative by the fundamental theorem of calculus as indicated in Fig. 1-12.

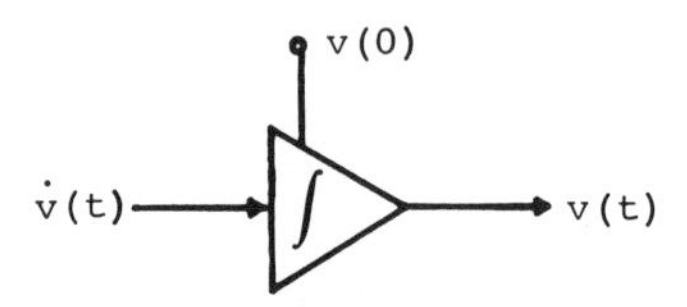

FIG. 1-12 A signal relationship involving an integrator.

The procedure for using these three elements to represent a state vector equation is analogous to the discrete-time case. The integrator is used to represent the relation between $x_i(t)$ and $\dot{x}_i(t)$, and these variables are interconnected according to the state vector equation using scalars and adders.

Example 9 An SVD for the single bucket system of Example 6 is shown in Fig. 1-13.

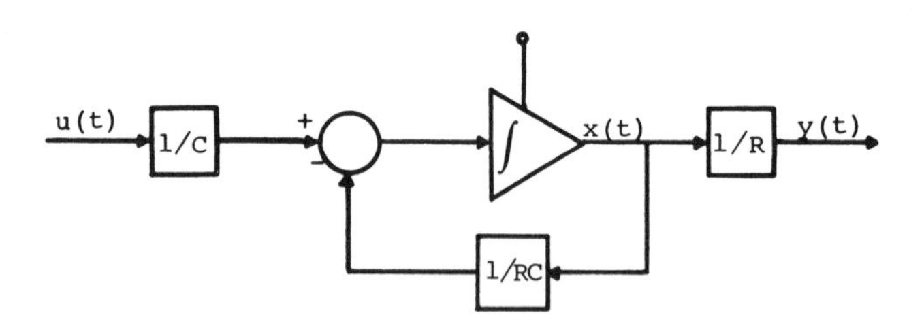

FIG. 1-13 A bucket system SVD.

Example 10 The circuit description (1.9) can be represented as shown in Fig. 1-14.

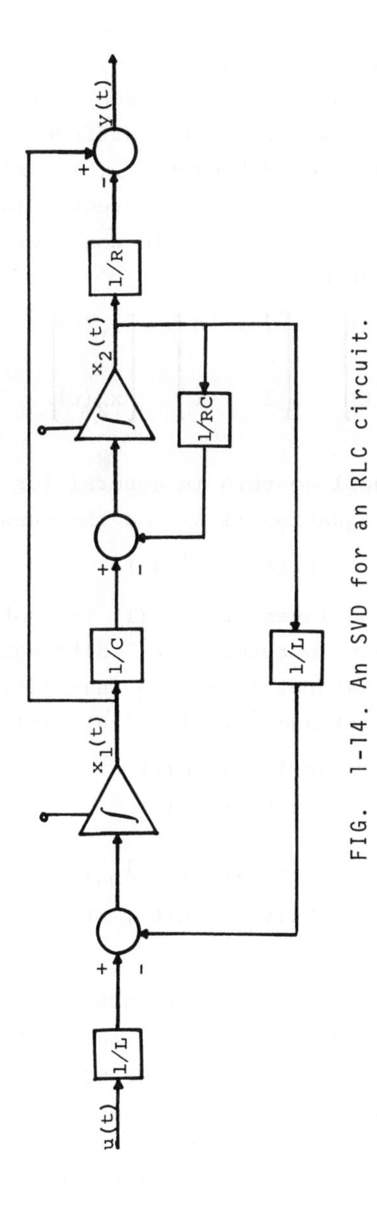

FIG. 1-14. An SVD for an RLC circuit.

1.5. CHANGE OF VARIABLES

It should be clear that there are many choices of state variables that can be used to obtain a state vector equation for a given system. For example, in the parallel bucket system, we could choose the depth of water in the first bucket $z_1(t) = x_1(t)$ and the difference in depth between the first and second buckets $z_2(t) = x_1(t) - x_2(t)$ and write a state vector equation directly in terms of these variables. Alternatively, we could consider the effect of the change of variables

$$\begin{bmatrix} z_1(t) \\ z_2(t) \end{bmatrix} = \begin{bmatrix} 1 & 0 \\ 1 & -1 \end{bmatrix} \begin{bmatrix} x_1(t) \\ x_2(t) \end{bmatrix}$$

on (1.8).

However, we shall do this in general for the continuous-time state vector equation (1.6) and the change of variable

$$z(t) = P^{-1}x(t) \tag{1.12}$$

where P is $n \times n$ and invertible. (It is customary to write this variable change in terms of P^{-1} although $z(t) = Px(t)$ might appear more natural.) Noting that $\dot{z}(t) = P^{-1}\dot{x}(t)$ and substituting for $x(t)$ and $\dot{x}(t)$ in (1.6) yields

$$\begin{aligned} P\dot{z}(t) &= APz(t) + bu(t) \\ y(t) &= cPz(t) + du(t) \end{aligned}$$

or

$$\begin{aligned} \dot{z}(t) &= P^{-1}APz(t) + P^{-1}bu(t), \quad z(0) = P^{-1}x_o \\ y(t) &= cPz(t) + du(t), \quad t \geq 0 \end{aligned} \tag{1.13}$$

Thus we obtain a new state vector equation in terms of $z(t)$, and this equation has precisely the same form as the original. We can use either state vector equation to describe the system since knowledge of $z(t)$ is equivalent to knowledge of $x(t)$.

The effect of this type of variable change in the discrete-time case is investigated in a similar manner.

The freedom to use different variables to describe the

internal structure of a system raises the possibility of choosing variables so that the coefficient matrices have particular structures. We will consider one such choice now. Two others will be considered in Chap. 2. It should be emphasized that these special variable choices are made for mathematical convenience and often the variables have no simple interpretation in terms of the physical system.

1.6. DISTINCT EIGENVALUE DIAGONAL FORM

Suppose we have a state vector equation of the form (1.6) where A has eigenvalues $\lambda_1, \ldots, \lambda_n$. Note that some of the λ_i may be complex numbers but that they must occur in complex conjugate pairs since A has real entries. Write the associated eigenvectors as $\delta_1, \ldots, \delta_n$. That is δ_i is a nonzero $n \times 1$ vector and $(A - \lambda_i I)\delta_i = 0$, $i = 1, \ldots, n$.

Consider the change of variables

$$z(t) = M^{-1}x(t) = \left[\delta_1 | \delta_2 | \ldots | \delta_n\right]^{-1} x(t)$$

To use this change of variables, we must verify that the so-called *modal* matrix M is indeed invertible.

Theorem 1 The modal matrix M is invertible if A has distinct eigenvalues.

Proof: We prove this by contradiction, that is, we assume that M is singular and use the assumption that A has distinct eigenvalues to arrive at an impossibility.

If M is singular then the columns of M must be linearly dependent and there exist scalars $a_2, a_3, \ldots, a_n$ such that

$$\delta_1 = a_2\delta_2 + a_3\delta_3 + \ldots + a_n\delta_n$$

We now proceed to multiply both sides of this equality on the left by the matrix $[(A - \lambda_n I)(A - \lambda_{n-1} I)\ldots(A - \lambda_2 I)]$ in a step-by-step fashion. Multiplying on the left by $(A - \lambda_2 I)$ and using the fact that each δ_i is an eigenvector of A yields

$$A\delta_1 - \lambda_2\delta_1 = a_2(A\delta_2 - \lambda_2\delta_2) + a_3(A\delta_3 - \lambda_2\delta_3) + \dots + a_n(A\delta_n - \lambda_2\delta_n)$$

or

$$(\lambda_1 - \lambda_2)\delta_1 = a_3(\lambda_3 - \lambda_2)\delta_3 + \dots + a_n(\lambda_n - \lambda_2)\delta_n$$

Multiplying both sides of this equality on the left by $(A - \lambda_3 I)$ yields

$$(\lambda_1 - \lambda_2)(\lambda_1 - \lambda_3)\delta_1 = a_4(\lambda_4 - \lambda_2)(\lambda_4 - \lambda_3)\delta_4 + \dots + a_n(\lambda_n - \lambda_2)(\lambda_n - \lambda_3)\delta_n$$

We continue by multiplying both sides of this equality on the left by $(A - \lambda_4 I)$. Finally, after repeating this calculation through the multiplication by $(A - \lambda_n I)$, we obtain

$$(\lambda_1 - \lambda_2)(\lambda_1 - \lambda_3)\dots(\lambda_1 - \lambda_n)\delta_1 = 0$$

But since the λ_i's are distinct, this implies that $\delta_1 = 0$. But this is a contradiction since δ_1 is an eigenvector and is thus nonzero. This completes the proof.

We can now compute the coefficient matrices of the new state vector equation under the assumption that A has distinct eigenvalues.

$$\begin{aligned}
M^{-1}AM &= [\delta_1|\delta_2|\dots|\delta_n]^{-1} A [\delta_1|\delta_2|\dots|\delta_n] \\
&= [\delta_1|\delta_2|\dots|\delta_n]^{-1} [A\delta_1|A\delta_2|\dots|A\delta_n] \\
&= [\delta_1|\delta_2|\dots|\delta_n]^{-1} [\lambda_1\delta_1|\lambda_2\delta_2|\dots|\lambda_n\delta_n] \\
&= [\delta_1|\delta_2|\dots|\delta_n]^{-1} [\delta_1|\delta_2|\dots|\delta_n] \begin{bmatrix} \lambda_1 & & & \\ & \lambda_2 & & \\ & & \ddots & \\ & & & \lambda_n \end{bmatrix} \\
&= \begin{bmatrix} \lambda_1 & & & \\ & \lambda_2 & & \\ & & \ddots & \\ & & & \lambda_n \end{bmatrix}
\end{aligned}$$

The coefficients cM and $M^{-1}b$ have no special structure. Thus the advantage of changing variables using the modal matrix is that $M^{-1}AM$ is diagonal. We remark again that in this case the coefficient matrix $M^{-1}AM$ may have complex elements although they must occur in complex conjugate pairs.

1.7. DISCRETE-TIME STATE VECTOR EQUATION SOLUTION

A general form for the solution of a discrete-time state vector equation with specified initial state and input sequence can be found by simple iteration as follows.

$$
\begin{aligned}
k = 0 \quad &: \quad x(1) = Ax_o + bu(0) \\
k = 1 \quad &: \quad x(2) = Ax(1) + bu(1) = A^2x_o + Abu(0) + bu(1) \\
&\qquad \vdots \\
k > 0 \quad &: \quad x(k) = A^kx_o + \sum_{j=0}^{k-1} A^{k-j-1}bu(j)
\end{aligned}
\tag{1.14}
$$

Thus the output sequence is given by

$$
y(k) = cA^kx_o + \sum_{j=0}^{k-1} cA^{k-j-1}bu(j) + du(k), \quad k > 0 \tag{1.15}
$$

Note that this gives the output as the sum of terms involving the initial state alone and terms involving the input sequence alone.

The above construction indicates that there is no difficulty with the existence of solutions to an arbitrary state vector equation of the form (1.1). A rather simple contradiction argument shows that this solution is unique. (See Problem 12)

Unfortunately the solution we have constructed is not in a simple closed form. That is, we do not have a simple analytical expression for the k-th term of the sequences $x(k)$ and $y(k)$. It is clear that the key to finding a closed form expression for $y(k)$ lies in finding a closed form expression for A^k , $k \geq 0$. For example, if A is diagonal we can write

$$A^k = \begin{bmatrix} \lambda_1 & & & \\ & \lambda_2 & & \\ & & \ddots & \\ & & & \lambda_n \end{bmatrix}^k = \begin{bmatrix} (\lambda_1)^k & & & \\ & (\lambda_2)^k & & \\ & & \ddots & \\ & & & (\lambda_n)^k \end{bmatrix}$$

However, note that if A is not diagonal but has distinct eigenvalues, then we can compute M so that

$$M^{-1}AM = \begin{bmatrix} \lambda_1 & & & \\ & \lambda_2 & & \\ & & \ddots & \\ & & & \lambda_n \end{bmatrix}$$

Then

$$(M^{-1}AM)^k = M^{-1}AM\ M^{-1}AM \ \dots\ M^{-1}AM$$
$$= M^{-1}A^kM$$

and we have that

$$A^k = M(M^{-1}AM)^kM^{-1} = M \begin{bmatrix} (\lambda_1)^k & & & \\ & (\lambda_2)^k & & \\ & & \ddots & \\ & & & (\lambda_n)^k \end{bmatrix} M^{-1} \qquad (1.16)$$

That is, for the case where A has distinct eigenvalues $\lambda_1, \dots, \lambda_n$, we can compute a closed form for A^k, and each element of A^k is a linear combination of the terms $(\lambda_1)^k, \dots, (\lambda_n)^k$. Of course these *exponentials* $(\lambda_i)^k$ may be complex but since A is real, we are guaranteed that the elements of A^k can be expressed as real numbers.

Substituting (1.16) into (1.15), we obtain a closed form for y(k) if u(k) can be expressed in a closed form. In later chapters we will discuss other means for obtaining closed form expressions.

Example 11 From Example 1 we know that the population distribution in the k-th generation of the Natchez Indians is given by

$$x(k) = \begin{bmatrix} 1 & 0 & 0 & 0 \\ 1 & 1 & 0 & 0 \\ 0 & 1 & 1 & 0 \\ -1 & -1 & 0 & 1 \end{bmatrix}^k x_o$$

It is easy to check that this A matrix has multiple eigenvalues so we will not attempt to compute A^k in closed form using the change of variables to diagonal form.

However, in this case there is a trick that can be used. We can write A = I + B where

$$B = \begin{bmatrix} 0 & 0 & 0 & 0 \\ 1 & 0 & 0 & 0 \\ 0 & 1 & 0 & 0 \\ -1 & -1 & 0 & 0 \end{bmatrix}$$

and by the binomial expansion (valid since I and B commute)

$$\begin{aligned} A^k &= (I + B)^k \\ &= I^k + \binom{k}{1} I^{k-1} B + \binom{k}{2} I^{k-2} B^2 + \ldots + \binom{k}{k-1} I B^{k-1} + B^k \end{aligned}$$

However $B^3 = 0$ and thus all higher powers of B are also zero. Therefore

$$\begin{aligned} A^k &= I + \binom{k}{1} B + \binom{k}{2} B^2 \\ &= \begin{bmatrix} 1 & 0 & 0 & 0 \\ k & 1 & 0 & 0 \\ \frac{k(k-1)}{2} & k & 1 & 0 \\ \frac{-k(k+1)}{2} & -k & 0 & 1 \end{bmatrix} \end{aligned}$$

and the population distribution is given by

$$x(k) = \begin{bmatrix} x_1(0) \\ kx_1(0) + x_2(0) \\ \frac{1}{2}k(k-1)x_1(0) + kx_2(0) + x_3(0) \\ -\frac{1}{2}k(k+1)x_1(0) - kx_2(0) + x_4(0) \end{bmatrix} \tag{1.17}$$

Note that no constant population distribution exists unless $x_1(0) = x_2(0) = 0$, that is, Suns and Nobles are initially absent from the population. But if this is so, the distribution remains at the initial distribution.

1.8. CONTINUOUS-TIME STATE VECTOR EQUATION SOLUTION

A general form for the solution of a continuous-time state vector equation with specified initial state and input signal can be found by an iterative procedure. However this procedure is more complicated and more delicate than that in the discrete-time case. In particular we need the following result; the proof of which is left to Problem 23.

<u>Theorem 2</u> If $F(t)$ is a matrix of integrable time functions defined for $t \geq 0$, then for $t \geq 0$,

$$\int_0^t \int_0^{\sigma_1} \cdots \int_0^{\sigma_{j-1}} F(\sigma_j)\, d\sigma_j\, d\sigma_{j-1} \cdots d\sigma_1 = \int_0^t F(\sigma_1) \frac{(t-\sigma_1)^{j-1}}{(j-1)!}\, d\sigma_1 \tag{1.18}$$

Beginning with the state vector equation

$$\dot{x}(t) = Ax(t) + bu(t),\ t \geq 0,\ x(0) = x_o$$

and the assumption that $u(t)$ is piecewise continuous, we can integrate both sides to obtain

$$x(t) = x_o + \int_0^t Ax(\sigma_1)\, d\sigma_1 + \int_0^t bu(\sigma_1)\, d\sigma_1, \quad t \geq 0 \tag{1.19}$$

Thus we can write

$$x(\sigma_1) = x_o + \int_0^{\sigma_1} Ax(\sigma_2)\, d\sigma_2 + \int_0^{\sigma_1} bu(\sigma_2)\, d\sigma_2, \quad \sigma_1 \geq 0$$

and substitute into (1.19) to obtain

$$x(t) = x_o + \int_0^t Ax_o\, d\sigma_1 + \int_0^t \int_0^{\sigma_1} A^2x(\sigma_2)\, d\sigma_2\, d\sigma_1$$

$$+ \int_0^t \int_0^{\sigma_1} Abu(\sigma_2)\, d\sigma_2\, d\sigma_1 + \int_0^t bu(\sigma_1)\, d\sigma_1$$

Integrating the second term and using (1.18) to rewrite the fourth term gives

$$x(t) = x_o + Atx_o + \int_0^t \int_0^{\sigma_1} A^2x(\sigma_2)\, d\sigma_2\, d\sigma_1 \tag{1.20}$$

$$+ \int_0^t Abu(\sigma_1)(t-\sigma_1)\, d\sigma_1 + \int_0^t bu(\sigma_1)\, d\sigma_1$$

Using (1.19) we can write

$$x(\sigma_2) = x_o + \int_0^{\sigma_2} Ax(\sigma_3)\, d\sigma_3 + \int_0^{\sigma_2} bu(\sigma_3)\, d\sigma_3$$

and substituting into (1.20) gives

$$x(t) = x_o + Atx_o + \int_0^t \int_0^{\sigma_1} A^2x_o\, d\sigma_2\, d\sigma_1$$

$$+ \int_0^t \int_0^{\sigma_1} \int_0^{\sigma_2} A^3x(\sigma_3)\, d\sigma_3\, d\sigma_2\, d\sigma_1 \tag{1.21}$$

$$+ \int_0^t \int_0^{\sigma_1} \int_0^{\sigma_2} A^2bu(\sigma_3)\, d\sigma_3\, d\sigma_2\, d\sigma_1$$

$$+ \int_0^t Abu(\sigma_1)(t-\sigma_1)\, d\sigma_1 + \int_0^t bu(\sigma_1)\, d\sigma_1$$

Integrating the third term in (1.21) and using (1.18) to rewrite the fifth term, we have

$$x(t) = x_o + Atx_o + \frac{1}{2}A^2t^2x_o + \int_0^t \int_0^{\sigma_1} \int_0^{\sigma_2} A^3x(\sigma_3)\, d\sigma_3\, d\sigma_2\, d\sigma_1$$

$$+ \int_0^t A^2bu(\sigma_1)\frac{1}{2}(t-\sigma_1)^2\, d\sigma_1 + \int_0^t Abu(\sigma_1)(t-\sigma_1)\, d\sigma_1$$

$$+ \int_0^t bu(\sigma_1)\, d\sigma_1$$

$$= (I + At + \frac{1}{2}A^2t^2)x_o$$

$$+ \int_0^t [I + A(t-\sigma_1) + \frac{1}{2}A^2(t-\sigma_1)^2]bu(\sigma_1)\, d\sigma_1$$

$$+ \int_0^t \int_0^{\sigma_1} \int_0^{\sigma_2} A^3x(\sigma_3)\, d\sigma_3\, d\sigma_2\, d\sigma_1$$

Continuing this process j times gives

$$x(t) = (I + \frac{1}{1!}At + \frac{1}{2!}A^2t^2 + \ldots + \frac{1}{j!}A^jt^j)x_o$$

$$+ \int_0^t [I + \frac{1}{1!}A(t-\sigma_1) + \frac{1}{2!}A^2(t-\sigma_1)^2 + \ldots + \frac{1}{j!}A^j(t-\sigma_1)^j]bu(\sigma_1)\, d\sigma_1$$

$$+ \int_0^t \int_0^{\sigma_1} \ldots \int_0^{\sigma_j} A^{j+1}x(\sigma_{j+1})d\sigma_{j+1}\, d\sigma_j \ldots d\sigma_1$$

We want to show that as $j \to \infty$ the series representation obtained for $x(t)$ converges uniformly in any finite interval $0 \le t \le T$. That is, we want to show that given $\varepsilon > 0$ there exists a J such that the norm of the term

$$\int_0^t \int_0^{\sigma_1} \ldots \int_0^{\sigma_j} A^{j+1}x(\sigma_{j+1})\, d\sigma_{j+1}\, d\sigma_j \ldots d\sigma_1$$

is less than ε for all $j > J$ and for all t, $0 \le t \le T$. The details of this computation will be omitted.

Thus we have obtained the infinite series representation

$$x(t) = \sum_{j=0}^{\infty} \frac{1}{j!}A^jt^jx_o + \int_0^t \sum_{j=0}^{\infty} \frac{1}{j!}A^j(t-\sigma_1)^jbu(\sigma_1)\, d\sigma_1$$

By analogy with the scalar case, we define the *matrix exponential* by

$$e^{At} = \sum_{j=0}^{\infty} \frac{1}{j!} A^j t^j$$

and write

$$x(t) = e^{At}x_o + \int_0^t e^{A(t-\sigma_1)} bu(\sigma_1)\, d\sigma_1, \quad t \geq 0 \qquad (1.22)$$

Also

$$y(t) = ce^{At}x_o + \int_0^t ce^{A(t-\sigma_1)} bu(\sigma_1)\, d\sigma_1 + du(t), \quad t \geq 0 \qquad (1.23)$$

Since e^{At} is given by a uniformly convergent series and term-by-term differentiation yields

$$\sum_{j=0}^{\infty} \frac{d}{dt} \frac{1}{j!} A^j t^j = \sum_{j=1}^{\infty} \frac{1}{(j-1)!} A^j t^{j-1} = A \sum_{k=0}^{\infty} \frac{1}{k!} A^k t^k$$

which is clearly a uniformly convergent series, we have

$$\frac{d}{dt} e^{At} = Ae^{At}$$

Using this property it is readily shown that (1.22) indeed satisfies the given state vector equation (See Problem 17). It can also be shown that this is the unique solution of the state vector equation.

There is one situation in which the matrix exponential e^{At} is particularly simple. If A is diagonal,

$$A = \begin{bmatrix} \lambda_1 & & & \\ & \lambda_2 & & \\ & & \ddots & \\ & & & \lambda_n \end{bmatrix}$$

then it is readily verified that the series above has the form

$$e^{At} = \begin{bmatrix} 1+\lambda_1 t + \frac{1}{2!}\lambda_1^2 t^2 \ldots & & & \\ & 1+\lambda_2 t + \frac{1}{2!}\lambda_2^2 t^2 + \ldots & & \\ & & \ddots & \\ & & & 1+\lambda_n t + \frac{1}{2!}\lambda_n^2 t^2 + \ldots \end{bmatrix}$$

$$= \begin{bmatrix} e^{\lambda_1 t} & & & \\ & e^{\lambda_2 t} & & \\ & & \ddots & \\ & & & e^{\lambda_n t} \end{bmatrix}$$

In fact this is the basis for one method of computing e^{At} in closed form. If A has distinct eigenvalues, then we can compute the modal matrix M so that

$$M^{-1}AM = \begin{bmatrix} \lambda_1 & & & \\ & \lambda_2 & & \\ & & \ddots & \\ & & & \lambda_n \end{bmatrix}$$

Then

$$e^{M^{-1}AMt} = I + (M^{-1}AM)t + \frac{1}{2!}(M^{-1}AM)^2 t^2 + \ldots + \frac{1}{j!}(M^{-1}AM)^j t^j + \ldots$$

$$= M^{-1}M + M^{-1}AMt + \frac{1}{2!}M^{-1}A^2 Mt^2 + \ldots + \frac{1}{j!}M^{-1}A^j Mt^j + \ldots$$

$$= M^{-1}(I + At + \frac{1}{2!}A^2 t^2 + \ldots + \frac{1}{j!}A^j t^j + \ldots)M$$

$$= M^{-1}e^{At}M,$$

and we can write

$$e^{At} = Me^{M^{-1}AMt}M^{-1} = M \begin{bmatrix} e^{\lambda_1 t} & & & \\ & e^{\lambda_2 t} & & \\ & & \ddots & \\ & & & e^{\lambda_n t} \end{bmatrix} M^{-1} \qquad (1.24)$$

From this expression we see that when A has distinct eigenvalues, each element of e^{At} is a linear combination $e^{\lambda_1 t}$, ..., $e^{\lambda_n t}$. As in the discrete-time case we know from the series definition that if A is real then e^{At} is real. Thus $e^{\lambda_i t}$ may be a complex exponential, but its complex conjugate must also appear so that the combinations can be written as real trigonometric functions of t.

Example 12 Consider the case

$$A = \begin{bmatrix} 0 & 1 \\ 0 & 0 \end{bmatrix}$$

which arises from the simple SVD shown in Fig. 1-15. Since

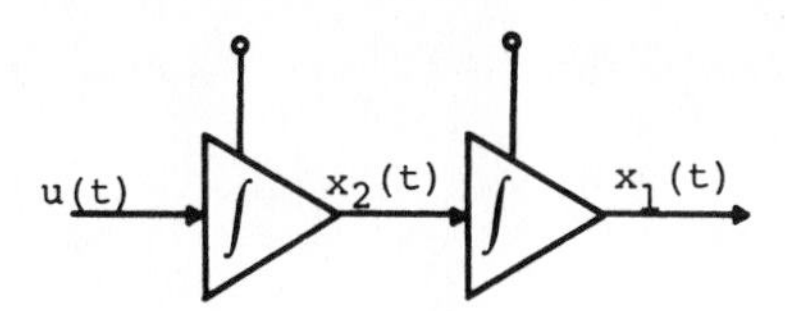

FIG. 1-15. A double integrator SVD.

$A^j = 0$ for $j \geq 2$ we have

$$e^{At} = I + At = \begin{bmatrix} 1 & t \\ 0 & 1 \end{bmatrix}$$

Example 13 Following Example 8 for the particular equation $\ddot{y}(t) + \omega^2 y(t) = 0$ yields the state vector equation

$$\dot{x}(t) = \begin{bmatrix} 0 & 1 \\ -\omega^2 & 0 \end{bmatrix} x(t), \quad t \geq 0$$

The eigenvalues of this A matrix are $\lambda_1 = i\omega$, $\lambda_2 = -i\omega$, and and the associated modal matrix is

$$M = \begin{bmatrix} 1 & 1 \\ i\omega & -i\omega \end{bmatrix}$$

Then

$$M^{-1}AM = \begin{bmatrix} i\omega & 0 \\ 0 & -i\omega \end{bmatrix}$$

and

$$e^{At} = M \begin{bmatrix} e^{i\omega t} & 0 \\ 0 & e^{-i\omega t} \end{bmatrix} M^{-1}$$

$$= \begin{bmatrix} \cos \omega t & \frac{1}{\omega} \sin \omega t \\ -\omega \sin \omega t & \cos \omega t \end{bmatrix}$$

Several properties of the matrix exponential are listed below for an $n \times n$ matrix A.

Property 1 e^{At} is a solution of the matrix differential equation

$$\dot{X}(t) = AX(t), \quad X(0) = I, \quad t \geq 0$$

Proof: Letting $X(t) = e^{At}$ we have from the series definition that $X(0) = I$ and $\dot{X}(t) = Ae^{At} = AX(t)$, $t \geq 0$.

Property 2 For any t_1 and t_2,

$$e^{At_1} e^{At_2} = e^{A(t_1+t_2)}$$

Proof: Since the defining series are absolutely convergent, we can compute $e^{At_1}e^{At_2}$ by term-by-term multiplication

$$e^{At_1}e^{At_2} = \sum_{k=0}^{\infty} \frac{1}{k!} A^k t_1^k \sum_{j=0}^{\infty} \frac{1}{j!} A^j t_2^j = \sum_{k=0}^{\infty} \sum_{j=0}^{\infty} \frac{1}{k!j!} A^{k+j} t_1^k t_2^j$$

Letting $i = k + j$, and using binomial coefficient properties,

$$e^{At_1}e^{At_2} = \sum_{i=0}^{\infty} A^i \frac{1}{i!} \sum_{j=0}^{i} \binom{i}{j} t_1^{i-j} t_2^j$$

$$= \sum_{i=0}^{\infty} \frac{1}{i!} A^i (t_1 + t_2)^i$$

$$= e^{A(t_1 + t_2)}$$

Property 3 e^{At} is invertible at any finite t and $[e^{At}]^{-1} = e^{-At}$.

Proof: To prove this property we need only show that

$$e^{At}\, e^{-At} = e^{-At}\, e^{At} = I$$

for this implies that e^{-At} is indeed the inverse of e^{At}. using term-by-term multiplication as in the proof of Property 2, we find

$$e^{At}\, e^{-At} = \sum_{k=0}^{\infty} \frac{1}{k!} A^k t^k \sum_{j=0}^{\infty} \frac{1}{j!} (-A)^j\, t^j$$

$$= \sum_{i=0}^{\infty} \frac{1}{i!} A^i\, (t-t)^i = I$$

The same result is obtained when $e^{-At}\, e^{At}$ is computed.

The following property also implies the invertibility of e^{At} for finite t. The proof is difficult and thus is omitted.

Property 4 For any t,

$$\det\, [e^{At}] = e^{(\text{trace } A)t}$$

Property 5 There exist n scalar functions $\alpha_0(t)$, $\alpha_1(t)$, ..., $\alpha_{n-1}(t)$ such that for all $t \geq 0$,

$$e^{At} = \sum_{i=0}^{n-1} \alpha_i(t) A^i \tag{1.25}$$

Proof: The Cayley-Hamilton Theorem states that if

$$\det(\lambda I - A) = \lambda^n + a_{n-1}\lambda^{n-1} + \dots + a_1\lambda + a_0$$

then

$$A^n + a_{n-1}A^{n-1} + \dots + a_1A + a_0I = 0$$

Thus A^n can be expressed as a linear combination of I, A, ..., A^{n-1},

$$A^n = -a_0I - a_1A - \dots - a_{n-1}A^{n-1} \tag{1.26}$$

Multiplying both sides by A and replacing the A^n term on the right side using (1.26) gives

$$A^{n+1} = (a_{n-1}a_0)I - (a_0 - a_{n-1}a_1)A - \dots - (a_{n-2} - a_{n-1}^2)A^{n-1} \tag{1.27}$$

Thus A^{n+1} can be expressed as a linear combination of I, A, ..., A^{n-1}. Similarly, multiplying both sides of (1.27) by A and using (1.26) to remove the A^n term on the right side shows that A^{n+2} can be expressed as a linear combination of I, A, ..., A^{n-1}.

This process clearly can be continued. Thus for each $j \geq 0$ there exists n scalars $a_{0,j}$, $a_{1,j}$, ..., $a_{n-1,j}$ such that

$$A^{n+j} = a_{0,j}I + a_{1,j}A + \dots + a_{n-1,j}A^{n-1}$$

Using this result in the series

$$e^{At} = I + At + \frac{1}{2!}A^2t^2 + \dots + \frac{1}{j!}A^jt^j + \dots$$

shows that the series can be written entirely in terms of I, A , ..., A^{n-1} and since the series converges absolutely, we can rearrange the various terms without changing the limit. Collecting together all the terms involving I, all those involving A, and so on, yields an expression of the form (1.25).

It is also worthwhile to note two properties that the

matrix exponential does *not* have. The i,j element of e^{At} is not $e^{a_{ij}t}$. In general, $e^{At}e^{Bt} \neq e^{(A+B)t}$, although a necessary and sufficient condition for equality is that A and B commute, AB = BA.

PROBLEMS

1. Draw a neat SVD corresponding to the state vector equation (1.5) in Example 3.
2. Using the same units and assumptions as Example 6, write a state vector equation for the "series" bucket system shown in Fig. 1-16.

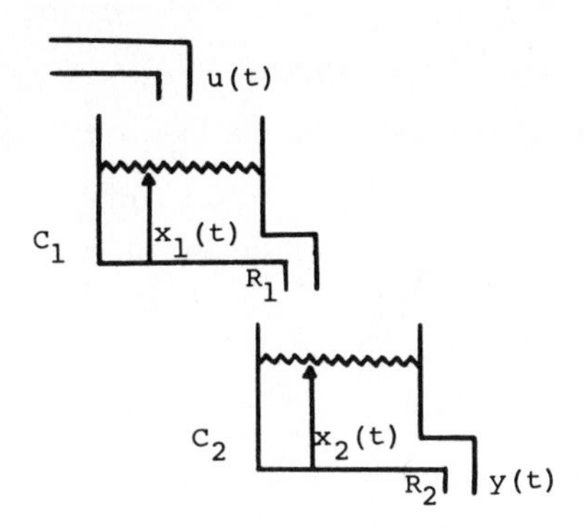

FIG. 1-16. Series water bucket system for Problem 2.

3. Write a state vector equation for the electrical circuit shown in Fig. 1-17 where u(t) is the input and the output is the current through R_2.
4. Draw a neat SVD for the parallel bucket system of Example 6.
5. Write a state vector equation for the system represented by the SVD shown in Fig. 1-18.
6. Define a set of state variables and write a state vector equation of the form (1.1) for the difference

equation

$$y(k+n) + a_{n-1}y(k+n-1) + \ldots + a_1y(k+1) + a_0y(k) = b_0u(k) + b_1u(k+1)$$

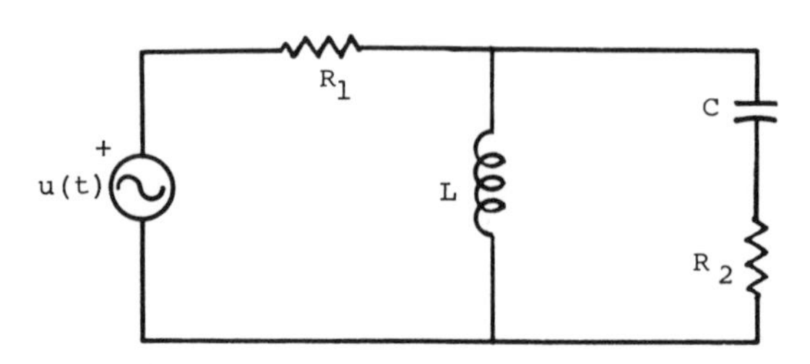

FIG. 1-17. RLC circuit for Problem 3.

7. Draw a neat SVD for the state vector equation from Problem 6.

8. Draw a general form for the SVD corresponding to a distinct eigenvalue diagonal form state vector equation.

9. Given the state vector equation (1.1) and a change of variables $z(k) = P^{-1}x(k)$, write a state vector equation in terms of z(k).

10. In Example 7 check that with L = 1, R = 1, and C = 1, the A-matrix has distinct eigenvalues, and find the diagonal form state vector equation.

11. Given a discrete-time state vector equation with A possessing distinct eigenvalues, devise a change of variables to obtain a diagonal form state vector equation.

12. Show that the solution of the discrete-time state vector equation given in (1.14) is unique.

13. Show that the output signal of a discrete-time state vector equation with specified input signal and zero initial state does not change when a change of variables $z(k) = P^{-1}x(k)$ is performed.

14. Assume a power series solution of $\dot{x}(t) = Ax(t)$ and show that the matrix exponential is obtained.

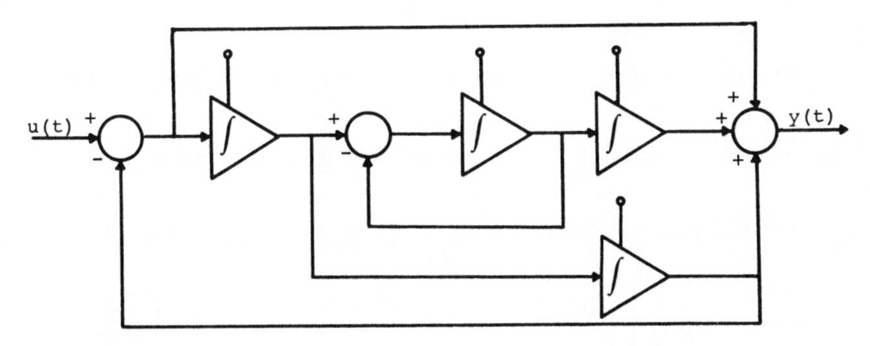

FIG. 1-18. SVD for Problem 5.

15. For the parallel bucket system in Example 6 with all constants unity, find a closed form expression for y(t) when x(0) = 0 and u(t) = 1 for all $t \geq 0$.

16. Show that if A is nonsingular, $\int_0^t e^{A\sigma} d\sigma = A^{-1}(e^{At} - I)$.

17. By differentiation of $x(t) = e^{At}x_o + \int_0^t e^{A(t-\sigma)} bu(\sigma)\,d\sigma$ verify that $\dot{x}(t) = Ax(t) + bu(t)$, $x(0) = x_o$.

18. Consider a continuous-time state vector equation with $x_o = 0$. Suppose that the input signal $u_1(t)$ yields the output signal $y_1(t)$ and $u_2(t)$ yields $y_2(t)$. Show that the input signal $u_1(t) + u_2(t)$ yields the output signal $y_1(t) + y_2(t)$ and the input signal $au_1(t)$ yields the output signal $ay_1(t)$. Formulate and prove a similar property for the response to initial states when the input is 0. Do similar properties hold in the discrete-time case?

19. Show that a solution of the $n \times n$ matrix equation $\dot{X}(t) = AX(t) + X(t)B$, $X(0) = X_o$, is $X(t) = e^{At}X_o e^{Bt}$.

20. For the continuous-time state vector equation (1.6), suppose that a *time variable* change of variables,

$$z(t) = P^{-1}(t)x(t)$$

is used. What is the form of the state vector equation in z(t)? What assumptions are needed on P(t)?

21. Following Problem 20, investigate the case where $P(t) = e^{At}$ and comment on any special features of the state vector equation in terms of the new state vector $z(t)$.

22. Show that the set of eigenvalues of A is identical to the set of eigenvalues of $P^{-1}AP$, where P is an arbitrary nonsingular matrix.

23. Prove Theorem 2.

24. Show that for continuous-time signals $f(t)$ and $g(t)$,
$$\int_0^t f(t-\sigma)g(\sigma)\, d\sigma = \int_0^t f(\sigma)g(t-\sigma)\, d\sigma.$$

REMARKS AND REFERENCES

1. We have not discussed the definitions of basic terms such as *system* and *linear*. Also we have not discussed the abstract notion of the *state* of a system. An exposition of these fundamental concepts can be found in the first chapter of L. A. Zadeh and E. Polak, System Theory, McGraw-Hill Book Co., New York, 1969.

2. We insist on using the standard form state vector equation description to facilitate the investigation of the structure of linear systems. However, the standard form also facilitates the use of a computer as a computational aid since vector/matrix calculations are easily programmed.

3. The Natchez Indian model is due to R. R. Bush. For further background material see C. W. M. Hart, "A Reconsideration of the Natchez Social Structure," American Anthropologist, New Series, 1943.

4. State variable diagrams can be formed by any interconnection of adders, scalars, and unit delayors (or integrators in the continuous-time case) as long as every loop contains at least one unit delayor (integrator). A loop which violates this rule may result in an indeterminate signal. For example in Fig. 1-19 $y(t)$ is not well defined when $a = -1$.

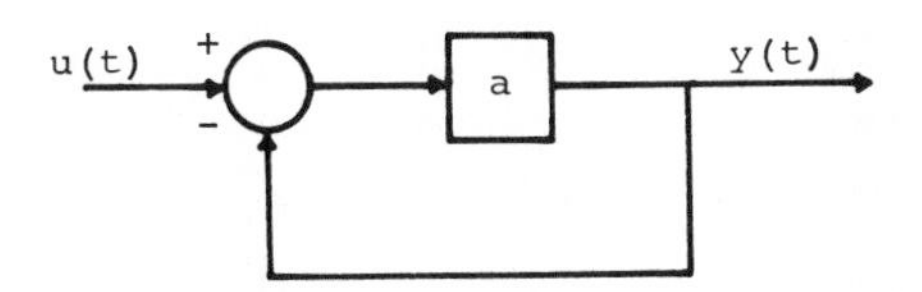

FIG. 1-19. An SVD which is ill defined when a = -1.

5. Throughout this text we will consider only single input/ single output systems. The generalization of the mathematical descriptions obtained here to the multi input/ multi output case is reasonably straightforward.

6. For the case where A has repeated eigenvalues, the diagonal form state vector equation cannot be obtained in general. However, a Jordan canonical form state vector equation can be found. See C. T. Chen, Introduction to Linear System Theory, Holt, Rhinehart, Winston, New York, 1970.

7. A rigorous development of the matrix exponential solution of the continuous-time state vector equation can be found in Chapter 5 of L. A. Zadeh and C. Desoer, Linear System Theory, McGraw-Hill Book Co., New York, 1963.

8. The state vector equation description of time-variable parameter linear systems is formulated in a manner similar to that given here. See R. Brockett, Finite Dimensional Linear Systems, John Wiley and Sons, New York, 1970. A proof of Property 4 of the matrix exponential is also given in this reference.

9. The state vector equation description of linear finite state systems (linear discrete-time systems over finite fields) is discussed in A. Gill, Linear Sequential Circuits, McGraw-Hill Book Co., New York, 1966. An advanced algebraic theory of linear systems which holds for arbitrary fields is developed in Chapter 10 of R. E. Kalman, P. L. Falb, M. A. Arbib, Topics in Mathematical System Theory, McGraw-Hill Book Co., New York, 1969.

10. Chap. 2 through Chap. 6 continue the study of state vector equations of the particular form (A,b,c,d). However, in Chap. 7 more general forms of state vector equations such as nonlinear equations will be discussed.

CHAPTER 2

COMPLETE REACHABILITY AND COMPLETE OBSERVABILITY

The state vector equation introduced in Chap. 1 makes evident the connections between the internal variables used to describe the system. In this chapter we introduce two fundamental concepts dealing with the interactions of the input and output with these variables. Roughly speaking, the property of complete reachability implies that the input signal can influence each state variable independently. Complete observability implies that the output signal is influenced by each state variable in an independent fashion. Thus if a state vector equation possesses these two properties there are no state variables which are disconnected from the input or output.

We also introduce two changes of variables which are intimately connected with these concepts. The canonical form state vector equations resulting from these variable changes will be used extensively in later chapters.

2.1. COMPLETE REACHABILITY IN THE CONTINUOUS-TIME CASE

We begin with a definition of complete reachability and a necessary and sufficient condition for a state vector equation to possess this property.

<u>Definition 1</u> The state vector equation

$$\begin{aligned} \dot{x}(t) &= Ax(t) + bu(t), \quad x(0) = x_o \\ y(t) &= cx(t) + du(t), \quad t \geq 0 \end{aligned} \tag{2.1}$$

is called *completely reachable* (CR) if for $x_o = 0$ and any desired state x_1 there exists a finite time t_1 and a piecewise continuous input $u(t)$, $0 \leq t \leq t_1$, such that $x(t_1) = x_1$.

In looser terminology, a state vector equation is CR if any desired state transfer from the zero initial state can be accomplished in finite time.

Example 1 Consider the water bucket system depicted in Fig. 2-1. In this system the depth of water in the first

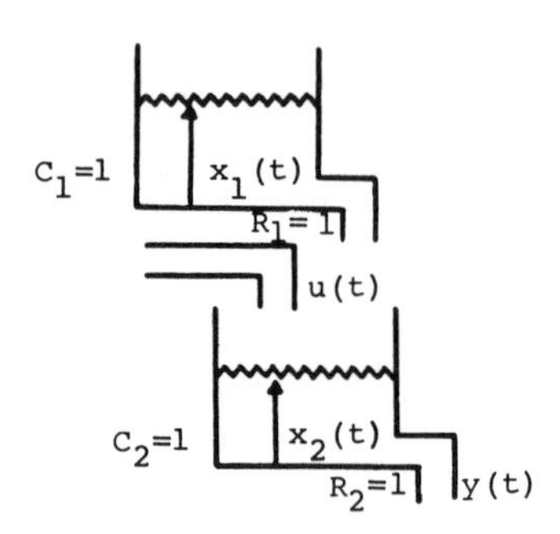

FIG. 2-1. A two bucket system.

tank $x_1(t)$ cannot be influenced by u(t). Thus we do not expect the state vector equation corresponding to this system to be CR.

It is of primary importance to determine conditions on the state vector equation (2.1) which guarantee the property of CR.

Theorem 1 The state vector equation (2.1) is CR iff

$$\text{rank}\left[b|Ab|\ldots|A^{n-1}b\right] = n \qquad (2.2)$$

that is, iff $\left[b|Ab|\ldots|A^{n-1}b\right]$ is an invertible matrix.

Proof: Suppose that (2.2) does not hold, that is,

$$\text{rank}\left[b|Ab|\ldots|A^{n-1}b\right] < n \qquad (2.3)$$

We will show this implies that (2.1) is not CR by exhibiting a state x_1 that cannot be attained from $x_o = 0$. Using the solution of the state vector equation and Property 5 of the matrix exponential from Chap. 1, we can write for any $t_1 > 0$ and $x_o = 0$,

$$x(t_1) = \int_0^{t_1} e^{A(t_1-\sigma)} bu(\sigma)\, d\sigma$$
$$= \sum_{i=0}^{n-1} \int_0^{t_1} \alpha_i(t_1-\sigma)u(\sigma)\, d\sigma\, A^i b \tag{2.4}$$

From (2.3) we have that the vectors b, Ab, ..., $A^{n-1}b$ are linearly dependent and thus we can pick an $n \times 1$ vector x_1 which is not in the span of these vectors. But (2.4) implies that every possible $x(t_1)$ is a linear combination of b, Ab, .., $A^{n-1}b$ and thus we cannot have $x(t_1) = x_1$.

Now suppose that (2.2) holds. We will show that the state vector equation is CR and, furthermore, that we may pick any $t_1 > 0$. That is, we will show that if a state vector equation satisfies (2.2), then for any given $t_1 > 0$ and x_1, there is an input $u(t)$, $0 \le t \le t_1$, such that $x(t_1) = x_1$.

First we will show that (2.2) implies the $n \times n$ matrix

$$\int_0^{t_1} e^{-A\sigma}bb'e^{-A'\sigma}\, d\sigma$$

is invertible for any $t_1 > 0$ by the method of contradiction. If there is a t_1 such that it is not invertible, then there is an $n \times 1$ vector $a \neq 0$ such that

$$a'\int_0^{t_1} e^{-A\sigma}bb'e^{-A'\sigma}\, d\sigma a = \int_0^{t_1} a'e^{-A\sigma}bb'e^{-A'\sigma}a\, d\sigma = 0$$

Since the integrand is the product of the two identical scalars, $a'e^{-A\sigma}b$ and $b'e^{-A'\sigma}a$, this implies that $a'e^{-A\sigma}b = 0$ for $0 \le \sigma \le t_1$. Thus

$$a'e^{-A\sigma}\, b\Big|_{\sigma=0} = a'b = 0$$

$$\frac{d}{d\sigma}\, a'e^{-A\sigma}\, b\Big|_{\sigma=0} = -a'Ae^{-A\sigma}\, b\Big|_{\sigma=0} = -a'Ab = 0$$

$$\vdots$$

$$\frac{d^{n-1}}{d\sigma^{n-1}}\, a'e^{-A\sigma}b\Big|_{\sigma=0} = (-1)^{n-1}a'A^{n-1}e^{-A\sigma}b\Big|_{\sigma=0}$$
$$= (-1)^{n-1}a'A^{n-1}b = 0$$

and we have that

$$a'A^i b = 0, \quad i = 0, \ 1, \ \ldots, \ n-1$$

That is

$$[a'b | a'Ab | \ldots | a' \ A^{n-1}b] = a'[b | Ab | \ldots | A^{n-1}b] = 0$$

which contradicts the assumption.

Using this result, given x_1 and t_1 we pick

$$u(t) = b'e^{-A't} \left[\int_0^{t_1} e^{-A\sigma}bb' \ e^{-A'\sigma} \ d\sigma\right]^{-1} e^{-At_1} x_1$$

Then from (2.4),

$$x(t_1) = \int_0^{t_1} e^{A(t_1-\sigma)} bb' \ e^{-A'\sigma} \left[\int_0^{t_1} e^{-A\sigma}bb' \ e^{-A'\sigma} d\sigma\right]^{-1} e^{-At_1} x_1 \ d\sigma$$

$$= e^{At_1} \int_0^{t_1} e^{-A\sigma}bb'e^{-A'\sigma} d\sigma \left[\int_0^{t_1} e^{-A\sigma}bb'e^{-A'\sigma} \ d\sigma\right]^{-1} e^{-At_1} x_1$$

$$= e^{At_1} \ e^{-At_1} \ x_1 = x_1$$

Thus with the chosen input we have $x(t_1) = x_1$.

<u>Example 2</u> The water bucket system of Example 1 is described by the state vector equation

$$\dot{x}(t) = \begin{bmatrix} -1 & 0 \\ 1 & -1 \end{bmatrix} x(t) + \begin{bmatrix} 0 \\ 1 \end{bmatrix} u(t)$$

$$y(t) = \begin{bmatrix} 0 & 1 \end{bmatrix} x(t)$$

Since

$$\text{rank } [b | Ab] = \text{rank} \begin{bmatrix} 0 & 0 \\ 1 & -1 \end{bmatrix} = 1$$

the conjecture of Example 1 is correct.

<u>Example 3</u> It is interesting to investigate the condition in (2.2) for the distinct eigenvalue diagonal form case,

$$\dot{x}(t) = \begin{bmatrix} \lambda_1 & & & \\ & \lambda_2 & & \\ & & \ddots & \\ & & & \lambda_n \end{bmatrix} x(t) + \begin{bmatrix} b_1 \\ b_2 \\ \vdots \\ b_n \end{bmatrix} u(t) \qquad (2.5)$$

It is clearly necessary for $b_i \neq 0$, $i = 1, 2, \ldots, n$ to have CR since $b_i = 0$ implies that $x_i(t)$ cannot be influenced by $u(t)$. To see if this is sufficient we check (2.2):

$$\text{rank} \begin{bmatrix} b_1 & \lambda_1 b_1 & \lambda_1^2 b_1 & \cdots & \lambda_1^{n-1} b_1 \\ b_2 & \lambda_2 b_2 & \lambda_2^2 b_2 & \cdots & \lambda_2^{n-1} b_2 \\ \vdots & \vdots & \vdots & & \vdots \\ b_n & \lambda_n b_n & \lambda_n^2 b_n & \cdots & \lambda_n^{n-1} b_n \end{bmatrix}$$

$$= \text{rank} \begin{bmatrix} b_1 & & & \\ & b_2 & & \\ & & \ddots & \\ & & & b_n \end{bmatrix} \begin{bmatrix} 1 & \lambda_1 & \lambda_1^2 & \cdots & \lambda_1^{n-1} \\ 1 & \lambda_2 & \lambda_2^2 & \cdots & \lambda_2^{n-1} \\ \vdots & \vdots & \vdots & & \vdots \\ 1 & \lambda_n & \lambda_n^2 & \cdots & \lambda_n^{n-1} \end{bmatrix}$$

The second matrix on the right side is a Vandermonde matrix and is nonsingular since the λ_i's are distinct. The diagonal matrix is nonsingular iff $b_i \neq 0$, $i = 1, 2, \ldots, n$. Thus $b_i \neq 0$, $i = 1, 2, \ldots, n$ is a necessary and sufficient condition for CR. Note that the property of CR is independent of the particular values of the distinct eigenvalues of A.

2.2. COMPLETE OBSERVABILITY IN THE CONTINUOUS-TIME CASE

Definition 2 The state vector equation (2.1) is called *completely observable* (CO) if for $u(t) = 0$ for all $t \geq 0$, there exists a finite time $t_1 > 0$ such that knowledge of $y(t)$, $0 \leq t \leq t_1$, suffices to determine the initial state x_o.

Example 4 Consider the water bucket system shown in Fig. 2-2.

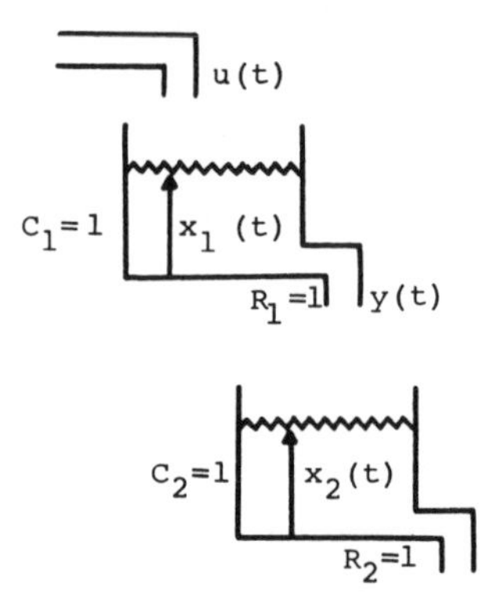

FIG. 2-2. Another two bucket system.

Since the output y(t) is not influenced by $x_2(t)$, we do not expect the state vector equation to be CO.

<u>Theorem 2</u> The state vector equation (2.1) is CO iff

$$\operatorname{rank}\begin{bmatrix} c \\ \hline cA \\ \hline \vdots \\ \hline cA^{n-1} \end{bmatrix} = n \tag{2.6}$$

that is iff the matrix is invertible.

<u>Proof</u>: Suppose that (2.6) does not hold. We will show that (2.1) is not CO by exhibiting an initial state $x_o \neq 0$ which yields the same output signal as the zero initial state, namely the zero output signal.

Since (2.6) does not hold we know there is an $n \times 1$ vector $x_o \neq 0$ such that

$$\begin{bmatrix} c \\ \hline cA \\ \hline \vdots \\ \hline cA^{n-1} \end{bmatrix} x_o = \begin{bmatrix} cx_o \\ \hline cAx_o \\ \hline \vdots \\ \hline cA^{n-1}x_o \end{bmatrix} = 0$$

Using Property 5 of the matrix exponential from Chap. 1, the output corresponding to $x(0) = x_o$ is given by

$$y(t) = ce^{At}x_o = c \sum_{i=0}^{n-1} \alpha_i(t) A^i x_o$$

$$= \sum_{i=0}^{n-1} \alpha_i(t)\ cA^i x_o$$

$$= 0, \quad \text{for all } t \geq 0$$

Since this is the same output as obtained from $x(0) = 0$, (2.1) is not CO.

Now suppose (2.6) holds. We will show that the state vector equation is CO and, furthermore, that we may pick any $t_1 > 0$ for the observation interval. Following the sufficiency proof of Theorem 1 we can show that if (2.6) holds then the $n \times n$ matrix

$$\int_0^{t_1} e^{A'\sigma}\ c'ce^{A\sigma}\ d\sigma$$

is invertible for any $t_1 > 0$.

For any x_o we can write

$$y(t) = ce^{At}x_o$$

and multiplying both sides by the column vector $e^{A't}c'$ gives

$$e^{A't}c'y(t) = e^{A't}c'ce^{At}x_o$$

Now given any $t_1 > 0$, we integrate both sides from 0 to t_1 to obtain

$$\int_0^{t_1} e^{A't}c'y(t)\ dt = \int_0^{t_1} e^{A't}c'ce^{At}\ dtx_o$$

Since the left side is known and the integral term on the right side is invertible, this equation yields a unique solution for x_o.

Example 5 The water bucket system of Example 4 is described by the state vector equation

$$\dot{x}(t) = \begin{bmatrix} -1 & 0 \\ 1 & -1 \end{bmatrix} x(t) + \begin{bmatrix} 1 \\ 0 \end{bmatrix} u(t)$$

$$y(t) = \begin{bmatrix} 1 & 0 \end{bmatrix} x(t)$$

As expected, this equation is not CO since

$$\text{rank} \begin{bmatrix} c \\ \hline cA \end{bmatrix} = \text{rank} \begin{bmatrix} 1 & 0 \\ -1 & 0 \end{bmatrix} = 1$$

2.3. COMPLETE REACHABILITY IN THE DISCRETE-TIME CASE

Definition 3 The state vector equation

$$\begin{aligned} x(k+1) &= Ax(k) + bu(k),\ x(0) = x_o \\ y(k) &= cx(k) + du(k),\ k = 0, 1, \ldots \end{aligned} \tag{2.7}$$

is called *completely reachable* (CR) if for the initial state $x(0) = 0$ and any desired state x_1 there is a finite positive integer k_1 and an input $u(k)$, $k = 0, 1, \ldots, k_1-1$ such that $x(k_1) = x_1$.

This definition is similar to Definition 1. It is perhaps surprising, and certainly interesting, that the conditions for CR are identical in the two cases.

Theorem 3 The state vector equation (2.7) is CR iff

$$\text{rank} \left[b | Ab | \ldots | A^{n-1}b \right] = n \tag{2.8}$$

Proof: Suppose that (2.8) holds. We will show that (2.7) is CR and, furthermore, that any desired transfer from $x_o = 0$ can be made with $k_1 = n$. Given x_1 we want to find scalars $u(0), u(1), \ldots, u(n-1)$ such that

$$x_1 = x(n) = \sum_{j=0}^{n-1} A^{n-j-1}\, bu(j) \tag{2.9}$$

But (2.8) implies that the vectors $b, Ab, \ldots, A^{n-1}b$ are linearly independent. Thus any $n \times 1$ vector x_1 can be expressed as in (2.9) and CR is assured.

Now suppose that (2.8) does not hold, that is, the rank

of $[b|Ab|\ldots|A^{n-1}b]$ is less than n. We will show that (2.7) is not CR by exhibiting an $n \times 1$ vector x_1 which cannot be attained from $x_o = 0$. The hypothesis implies that the vectors b, Ab, ..., $A^{n-1}b$ are linearly dependent and thus there is an $n \times 1$ vector x_1 which cannot be expressed as a linear combination of these vectors. But this implies that x_1 cannot be expressed as a linear combination of b, Ab, ..., $A^m b$ for any $m \geq 0$ since the Cayley-Hamilton Theorem can be used to express any vector $A^j b$ as a linear combination of b, Ab, ..., $A^{n-1}b$. Thus, regardless of m, an expression of the form

$$x_1 = x(m) = \sum_{j=0}^{m-1} A^{m-j-1} bu(j)$$

cannot hold and we have that the state vector equation is not CR.

<u>Example 6</u> In Chap. 1 the state vector equation description of a fish hatchery was found to be

$$x(k+1) = \begin{bmatrix} -a_3 & -a_2 & a_1 \\ 1 & -a_4 & 0 \\ 0 & a_5 & 1-a_6 \end{bmatrix} x(k) + \begin{bmatrix} 1 \\ 0 \\ 0 \end{bmatrix} u(k) \tag{2.10}$$

$$y(k) = [0 \quad a_4 \quad 0]\, x(k)$$

In this case

$$\operatorname{rank}\,[b|Ab|A^2b] = \operatorname{rank} \begin{bmatrix} 1 & -a_3 & a_3^2-a_2 \\ 0 & 1 & -a_3-a_4 \\ 0 & 0 & a_5 \end{bmatrix}$$

$$= \begin{cases} 3, & a_5 \neq 0 \\ 2, & a_5 = 0 \end{cases}$$

Thus (2.10) is CR iff $a_5 \neq 0$. This conclusion is plausible by inspection of the SVD corresponding to (2.10). Also from the system description in Chap. 1, Example 2, it is clear that if $a_5 = 0$, the input has no influence on $x_3(k)$.

The analysis of CR for the distinct eigenvalue diagonal form case is essentially the same as in Example 3.

2.4. COMPLETE OBSERVABILITY IN THE DISCRETE-TIME CASE

Definition 4 The state vector equation (2.7) is called *completely observable* (CO) if for $u(k) = 0$ for all $k \geq 0$, there is a finite positive integer k_1 such that knowledge of $y(0), y(1), \ldots, y(k_1-1)$ suffices to determine the initial state x_o.

Again, this definition is similar to that in the continuous-time case and the conditions for CO are identical.

Theorem 4 The state vector equation (2.7) is CO iff

$$\operatorname{rank}\begin{bmatrix} c \\ \hline cA \\ \hline \vdots \\ \hline cA^{n-1} \end{bmatrix} = n \tag{2.11}$$

Proof: Suppose that (2.11) holds. We will show that x_o can be determined from a knowledge of $y(0), y(1), \ldots, y(n-1)$. This implies that (2.7) is CO and that we can always take $k_1 = n$.

Since $u(k) = 0$ for all k, we have that $y(k) = cA^k x_o$, $k = 0, 1, \ldots, n-1$. Arranging this in vector form gives

$$\begin{bmatrix} y(0) \\ y(1) \\ \vdots \\ y(n-1) \end{bmatrix} = \begin{bmatrix} cx_o \\ \hline cAx_o \\ \hline \vdots \\ \hline cA^{n-1}x_o \end{bmatrix} = \begin{bmatrix} c \\ \hline cA \\ \hline \vdots \\ \hline cA^{n-1} \end{bmatrix} x_o$$

Since the vector of output values is assumed known and the matrix on the right side is invertible by (2.11), x_o is given by matrix inversion.

Now suppose that (2.11) does not hold. We will show

that the state vector equation is not CO by exhibiting an initial state $x_o \neq 0$ which yields the same output signal as $x(0) = 0$. The hypothesis implies that there is a nonzero $n \times 1$ vector x_o such that

$$0 = \begin{bmatrix} c \\ \hline cA \\ \hline \vdots \\ \hline cA^{n-1} \end{bmatrix} x_o = \begin{bmatrix} cx_o \\ \hline cAx_o \\ \hline \vdots \\ \hline cA^{n-1}x_o \end{bmatrix} = \begin{bmatrix} y(0) \\ y(1) \\ \vdots \\ y(n-1) \end{bmatrix}$$

Thus $y(k) = 0$, $k = 0, 1, \ldots, n-1$. For any $m > n-1$ we can use the Cayley-Hamilton Theorem to obtain

$$\begin{aligned} y(m) &= cA^m x_o = c[a_{0,m}I + a_{1,m}A + \ldots + a_{n-1,m}A^{n-1}]x_o \\ &= a_{0,m}cx_o + a_{1,m}cAx_o + \ldots + a_{n-1,m}cA^{n-1}x_o = 0 \end{aligned}$$

which is identical to the output produced by $x_o = 0$.

Example 7 Checking CO for the state vector equation of a fish hatchery in Example 6 with the assumption that $a_4 \neq 0$ gives

$$\text{rank} \begin{bmatrix} c \\ \hline cA \\ \hline cA^2 \end{bmatrix} = \text{rank} \begin{bmatrix} 0 & a_4 & 0 \\ a_4 & -a_4^2 & 0 \\ -a_3a_4-a_4^2 & -a_2a_4+a_4^3 & a_4a_1 \end{bmatrix} = \begin{cases} 3, & a_1 \neq 0 \\ 2, & a_1 = 0 \end{cases}$$

Thus the state vector equation is CO iff $a_1 \neq 0$. From the corresponding SVD or from the system description it is clear that if $a_1 = 0$, then the number of young fish (which determines the output) is independent of the number of adult fish $x_3(k)$.

2.5. REACHABILITY AND OBSERVABILITY CANONICAL FORMS

A state vector equation in which the coefficient matrices have the form

$$\dot{x}(t) = \begin{bmatrix} 0 & 1 & 0 & \cdots & 0 & 0 \\ 0 & 0 & 1 & \cdots & 0 & 0 \\ \vdots & \vdots & \vdots & & \vdots & \vdots \\ 0 & 0 & 0 & \cdots & 0 & 1 \\ -a_0 & -a_1 & -a_2 & \cdots & -a_{n-2} & -a_{n-1} \end{bmatrix} x(t) + \begin{bmatrix} 0 \\ 0 \\ \vdots \\ 0 \\ 1 \end{bmatrix} u(t)$$

$$y(t) = \begin{bmatrix} c_0 & c_1 & \cdots & c_{n-1} \end{bmatrix} x(t) + du(t) \qquad (2.12)$$

is said to be in *reachability canonical form* (RCF). Writing out the scalar equations in (2.12), it is easy to see that the corresponding SVD can be drawn as shown in Fig. 2-3.

As suggested by the terminology, RCF state vector equations have a special connection with the property of CR. We first show that an RCF equation is CR regardless of the values of the coefficients a_i and c_i, $i = 0, 1, \ldots, n-1$. Checking the CR condition gives

$$\operatorname{rank} \left[b \,|\, Ab \,|\, \ldots \,|\, A^{n-1}b \right] = \operatorname{rank} \begin{bmatrix} 0 & 0 & & 1 \\ 0 & 0 & & -a_{n-1} \\ \vdots & \vdots & .\cdot^{\cdot} & \vdots \\ 0 & 1 & \cdots & \\ 1 & -a_{n-1} & \cdots & \end{bmatrix} = n$$

since the matrix is lower triangular with nonzero antidiagonal elements. This result is made more intuitive by the inspection of the SVD where it is clear that, regardless of the scalars, each state variable is connected at least indirectly with the input.

Secondly we show that given a CR state vector equation (A,b,c,d), there is a nonsingular change of variables R such that $(R^{-1}AR, R^{-1}b, cR, d)$ is in RCF. To explicitly compute R, let

$$\det (\lambda I - A) = \lambda^n + a_{n-1}\lambda^{n-1} + \ldots + a_1\lambda + a_0 \qquad (2.13)$$

and define a set of $n \times 1$ vectors by

$$\begin{aligned} r_0 &= b \\ r_i &= Ar_{i-1} + a_{n-i}r_0, \quad i = 1, \ldots, n \end{aligned} \qquad (2.14)$$

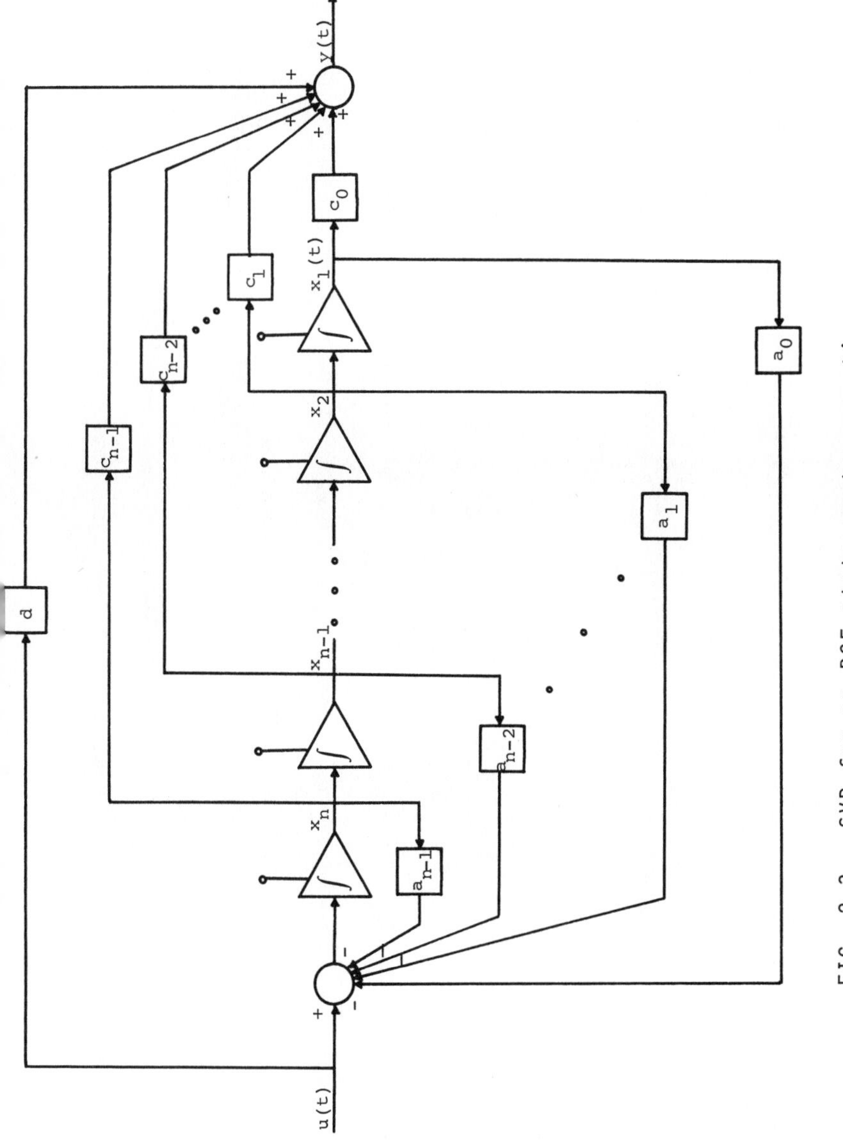

FIG. 2-3. SVD for an RCF state vector equation.

Since the vectors b, Ab, ..., $A^{n-1}b$ are linearly independent and since r_i is a linear combination of $A^i b$, $A^{i-1}b$, ..., b, i = 0, ..., n-1 it is readily seen that r_0, r_1, ..., r_{n-1} is a linearly independent set. Furthermore

$$\begin{aligned}
r_n &= Ar_{n-1} + a_0 r_0 \\
&= A^2 r_{n-2} + a_1 A r_0 + a_0 r_0 \\
&= A^3 r_{n-3} + a_2 A^2 r_0 + a_1 A r_0 + a_0 r_0 \\
&\vdots \\
&= A^n r_0 + a_{n-1} A^{n-1} r_0 + \dots + a_1 A r_0 + a_0 r_0 \\
&= 0
\end{aligned}$$

by the Cayley-Hamilton Theorem.

Now let

$$R = \left[r_{n-1} \middle| r_{n-2} \middle| \cdots \middle| r_0 \right] \tag{2.15}$$

Then R is invertible since the columns are linearly independent. We can verify that

$$R^{-1}b = \begin{bmatrix} 0 \\ 0 \\ \vdots \\ 0 \\ 1 \end{bmatrix}$$

by a simple calculation:

$$R \begin{bmatrix} 0 \\ 0 \\ \vdots \\ 0 \\ 1 \end{bmatrix} = \left[r_{n-1} \middle| r_{n-2} \middle| \cdots \middle| r_0 \right] \begin{bmatrix} 0 \\ 0 \\ \vdots \\ 0 \\ 1 \end{bmatrix} = r_0 = b$$

Similarly, to verify that $R^{-1}AR$ has the desired form, we compute

$$R\begin{bmatrix} 0 & 1 & \cdots & 0 \\ 0 & 0 & \cdots & 0 \\ \vdots & \vdots & & \vdots \\ 0 & 0 & \cdots & 1 \\ -a_0 & -a_1 & \cdots & -a_{n-1} \end{bmatrix} = \left[-a_0 r_0 \middle| r_{n-1} - a_1 r_0 \middle| \cdots \middle| r_1 - a_{n-1} r_0\right]$$

$$= \left[Ar_{n-1} \middle| Ar_{n-2} \middle| \cdots \middle| Ar_0\right] = AR$$

Thus we have the required variable change. Note that the bottom row elements of $R^{-1}AR$ are just the coefficients of the characteristic polynomial of A. These results are for both the continuous and discrete-time cases.

A state vector equation of the form

$$\dot{x}(t) = \begin{bmatrix} 0 & 0 & \cdots & 0 & -a_0 \\ 1 & 0 & \cdots & 0 & -a_1 \\ 0 & 1 & \cdots & 0 & -a_2 \\ \vdots & \vdots & & \vdots & \vdots \\ 0 & 0 & \cdots & 0 & -a_{n-2} \\ 0 & 0 & \cdots & 1 & -a_{n-1} \end{bmatrix} x(t) + \begin{bmatrix} b_0 \\ b_1 \\ b_2 \\ \vdots \\ b_{n-2} \\ b_{n-1} \end{bmatrix} u(t) \qquad (2.16)$$

$$y(t) = \begin{bmatrix} 0 & 0 & \cdots & 0 & 1 \end{bmatrix} x(t) + du(t)$$

is said to be in *observability canonical form* (OCF). Checking the CO condition it is readily shown that (2.16) is CO regardless of the parameter values. We can also show that for any CO state vector equation (A,b,c,d), there is a nonsingular change of variables Q such that $(Q^{-1}AQ, Q^{-1}b, cQ, d)$ is in OCF. In fact if (2.13) is the characteristic polynomial of A and if we define a set of $1 \times n$ vectors by

$$q_0 = c$$
$$q_i = q_{i-1}A + a_{n-i}q_0, \quad i = 1, \ldots, n \qquad (2.17)$$

then we take

$$Q^{-1} = \begin{bmatrix} q_{n-1} \\ \hline q_{n-2} \\ \hline \vdots \\ \hline q_0 \end{bmatrix} \tag{2.18}$$

The details are left as an exercise.

PROBLEMS

1. Consider the parallel bucket system shown in Fig. 2-4 where all parameter values are unity.
   a. If u(t) is applied to the first tank is the state vector equation CR?
   b. If the input is applied to the second tank is the state vector equation CR?

   Can you intuitively justify your answers?

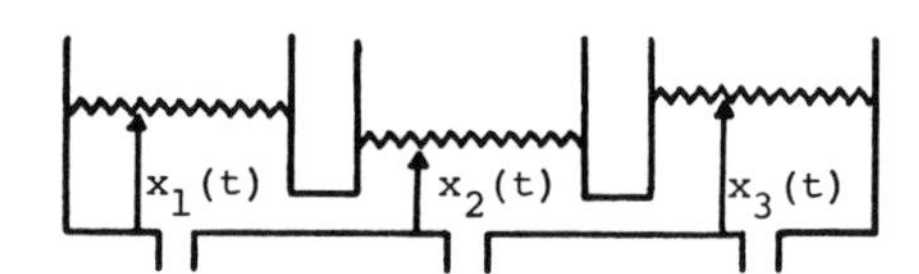

FIG. 2-4. Three buckets in parallel.

2. Devise a simple bucket system such that the state vector equation is neither CR nor CO.

3. Show that for a CO state vector equation with a known input u(t), x_o can be determined from a knowledge of y(t) for $0 \leq t \leq t_1$, $t_1 < \infty$.

4. For the bucket system of Problem 1, let the output be the outflow from the first tank. Is the state vector equation CO?

5. Are the state vector equations for the systems shown in Fig. 2-5 and 2-6 CO and CR?

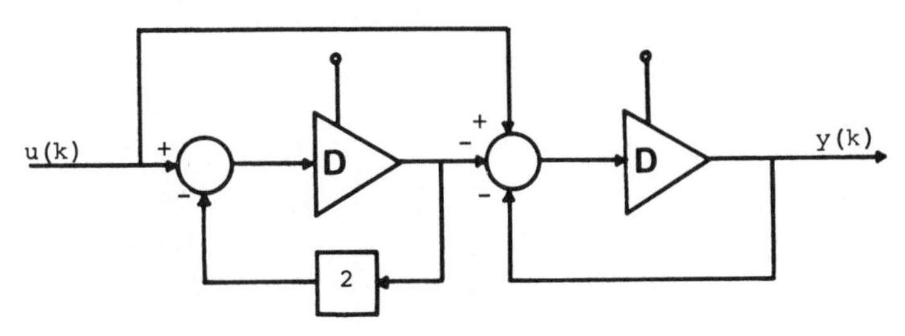

FIG. 2-5. SVD for Problem 5.

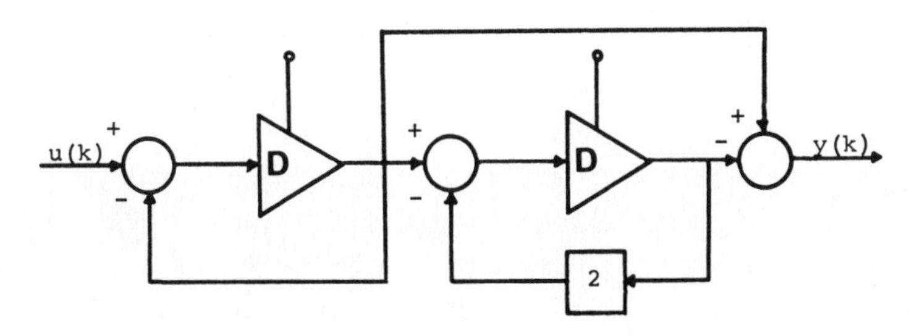

FIG. 2-6. SVD for Problem 5.

6. Investigate CO in the distinct eigenvalue diagonal form case.

7. In the continuous-time case show that the condition

$$\operatorname{rank}\begin{bmatrix} c \\ \hline cA \\ \hline \vdots \\ \hline cA^{n-1} \end{bmatrix} = n$$

implies CO by considering $y(t_1)$, $y^{(1)}(t_1)$, ..., $y^{(n-1)}(t_1)$ for any $t_1 \geq 0$.

8. Show that if two state vector equations are related by a nonsingular change of variables, then one is CR iff the other is CR. Show the same for CO.

9. Show that if a continuous-time state vector equation is CR then *any* initial state x_o can be transferred to any desired state x_1 in a finite time.

10. A discrete-time state vector equation is called *completely controllable* (CC) if for any initial state $x(0) = x_o$ there is a positive integer $k_1 < \infty$ and an input $u(k)$, $k = 0, 1, \ldots, k_1-1$ such that $x(k_1) = 0$. Find a state vector equation which is CC but not CR. Can you find a condition under which a state vector equation is CC iff it is CR?

11. Can you find an output for the Natchez Indian state vector equation which involves at most two subpopulations and is such that the state vector equation is CO?

12. A discrete-time state vector equation is called *completely detectable* (CD) if for $u(k) = 0$ for all k, and $k_1 > n$, a knowledge of $y(0), y(1), \ldots, y(k_1-1)$, $k_1 < \infty$ suffices to determine $x(k_1)$. Prove or find a counter example to the claim that a state vector equation is CD if it is CO.

13. Show that the state vector equation (A,b,c,d) is CR (CO) iff the state vector equation (A',c',b',d) is CO (CR).

14. Verify that Q defined in (2.17) and (2.18) transforms a CO state vector equation to OCF.

15. Give proofs or counterexamples to the following statements.
    a. If (A,b,c,d) is CR, then $(A-bh',b,c,d)$ is CR for all $n \times 1$ vectors h.
    b. If (A,b,c,d) is CO, then $(A-bh',b,c,d)$ is CO for all $n \times 1$ vectors h.

REMARKS AND REFERENCES

1. The terminology introduced in this chapter is not completely standard in the literature. For example the distinction between complete reachability (Definitions 1 and 3) and complete controllability (Problem 10) is often disregarded by imposing assumptions or using different definitions. Also reachability canonical form is often called *phase-variable* or *rational canonical* form.

2. In Definition 1 the set of admissable controls is taken to be the set of piecewise continuous functions of t. This restriction can be considerably weakened without changing the results. In fact all that is needed is that the integrals used in the proof of the theorem be well-defined.

3. We have defined the concepts of CR and CO for state vector equations and not for systems. This approach is taken because there are often system constraints that are not reflected in the mathematical description. For example in the bucket systems the variables $x_i(t)$, $u(t)$, and $y(t)$ must always be nonnegative. Thus in, say, Problem 1 we may conclude CR from the state vector equation description, but since negative input signals cannot be produced there may be desired states which cannot be reached. We take the view that this difficulty is not with the theory but in this particular application. (A more satisfactory framework for bucket systems will appear when linearization about operating points is discussed in Chap. 7.)

4. The definitions of CR and CO presented in this chapter have clear operational interpretations in terms of transferring the state or determining the initial state. However the concepts of CR and CO also play central structural roles in the relationship between internal and input/output system descriptions. These structural aspects will be considered in Chap. 4.

CHAPTER 3

EXTERNAL SYSTEM DESCRIPTION: INPUT/OUTPUT MAPS

Under the assumption of zero initial state, the state vector equation can be viewed as producing an output signal corresponding to a given input signal. The mathematical representation of this operation is called an input/output map. Such a representation is often used because detailed information about the internal structure of the system as represented by the state vector equation is not of interest or not available.

In this chapter we discuss input/output maps which arise from state vector equations of the form treated in Chaps. 1 and 2. The first input/output map is based on the state vector equation solution derived in Chap. 1. Then the transform representation for signals is introduced and several properties of this representation derived. Using the transform representation for input and output signals leads to the well known transfer function input/output map and the block diagram representation. The transform representation is also used in the development of more closed form expressions for matrix exponentials and matrix powers.

3.1. THE UNIT PULSE RESPONSE

For the state vector equation

$$\begin{aligned} x(k+1) &= Ax(k) + bu(k) \\ y(k) &= cx(k) + du(k), \quad k = 0, 1, 2, \ldots \end{aligned} \tag{3.1}$$

with $x(0) = 0$, the output can be expressed in terms of the input signal by

$$y(k) = \sum_{j=0}^{k-1} cA^{k-j-1}bu(j) + du(k)$$
$$= \sum_{j=0}^{k} h(k-j)u(j) \qquad (3.2)$$

where

$$h(k) = \begin{cases} d, & k = 0 \\ cA^{k-1}b, & k > 0 \end{cases} \qquad (3.3)$$

Thus if the scalar signal h(k) is known, the output signal can be computed for any input signal.

Suppose that the input is the *unit pulse* signal $u_0(k)$ defined by

$$u_0(k) = \begin{cases} 1, & k = 0 \\ 0, & k > 0 \end{cases} \qquad (3.4)$$

Substitution into (3.2) shows that the output is precisely the signal h(k). Thus if the unit pulse response is known, the response to any input signal can be computed. We therefore view the unit pulse response as specifying an input/output map. Of course it is desirable to have a closed form expression for h(k).

3.2. THE Z-TRANSFORM

The z-transform representation (see Table 3-1, at end of chapter) involves assigning to a discrete-time signal f(k) a unique function F(z) of the indeterminate z. We then show that common operations on discrete-time signals correspond to simple operations on F(z). Thus it is often more convenient to use the z-transform representation than the infinite sequence.

In addition to writing out the infinite sequence,

$$f(k) = (f(0), f(1), f(2), \ldots) \qquad (3.5)$$

we can write a signal as a sum of terms as long as a place marker is used to indicate which term in the sum corresponds to the k-th position in the sequence. We let z^{-k} be the place marker for the k-th position, $k \geq 0$. Then we can write

(3.5) in the form

$$f(k) = f(0)z^0 + f(1)z^{-1} + f(2)z^{-2} + \ldots \qquad (3.6)$$

or

$$f(k) = \sum_{i=0}^{\infty} f(i)z^{-i} \qquad (3.7)$$

We call (3.7) a *formal power series* since we do not consider the question of convergence. The use of the summation sign in this context does not indicate a numerical total since z is not a real or complex variable. The symbol z is used simply as a place marker which allows a sequence (3.5) to be written in the notation (3.7).

The advantage of the formal series representation is that under certain conditions we can find a rational function F(z) which when expanded yields the series. This function is sometimes called the generating function of the formal series. However, we will call F(z) the z-transform of f(k) and write F(z) = Z[f(k)].

The key mathematical tools involve Hankel matrices and rational functions.

Definition 1 The *Hankel matrix* H_f associated with a discrete-time signal f(k) is the symmetric matrix with an infinite number of rows and columns given by

$$H_f = \begin{bmatrix} f(1) & f(2) & f(3) & \ldots \\ f(2) & f(3) & f(4) & \ldots \\ f(3) & f(4) & f(5) & \ldots \\ \vdots & \vdots & \vdots & \end{bmatrix} \qquad (3.8)$$

Definition 2 A *rational function* of an indeterminate z is a ratio of polynomials in z. It is called *proper* if the degree of the numerator polynomial is no greater than the degree of the denominator polynomial and *strictly proper* if the numerator degree is less than the denominator degree. We call a rational function *reduced* if the numerator and denominator polynomials have no factors in common, and we call the degree of the denominator polynomial the *degree* of a rational function.

We will write a proper rational function of degree n in the standard form

$$F(z) = \frac{b_n z^n + b_{n-1} z^{n-1} + \ldots + b_1 z + b_0}{z^n + a_{n-1} z^{n-1} + \ldots + a_1 z + a_0} \qquad (3.9)$$

Recall that by ordinary long division of the numerator by the denominator, we can expand F(z) into a formal power series expression of the form (3.7). Of course the usual operations carry through the long division process. Scalar multiplication of F(z) corresponds to scalar multiplication of each term in (3.7) and addition of proper rational functions corresponds to term-by-term addition of the formal series coefficients. Also recall that multiplication of proper rational functions carries through as follows:

$$\begin{aligned} F(z)G(z) &= (f(0)+f(1)z^{-1}+f(2)z^{-2}+\ldots)(g(0)+g(1)z^{-1}+g(2)z^{-2}+\ldots) \\ &= f(0)g(0) + (f(0)g(1) + f(1)g(0))z^{-1} \\ &\quad + (f(0)g(2) + f(1)g(1) + f(2)g(0))z^{-2} \\ &\quad + \ldots \end{aligned}$$

<u>Lemma 1</u> For any discrete-time signal f(k), H_f has finite rank n iff there exist n scalars $a_0, a_1, \ldots, a_{n-1}$ such that

$$f(k+1) = \sum_{q=0}^{n-1} a_q f(k-q), \quad k = n, n+1, \ldots \qquad (3.10)$$

and n is the least number having this property.

<u>Proof</u>: We first note that the case n = 0 is trivial. If H_f has finite rank n > 0, then the first n+1 rows, say, $R_1, R_2, \ldots, R_{n+1}$, are linearly dependent. Then there is an integer $1 \le p \le n$ such that $R_1, R_2, \ldots, R_p$ are linearly independent and R_{p+1} can be expressed as a linear combination of them,

$$R_{p+1} = \sum_{q=0}^{p-1} a_q R_{p-q} \tag{3.11}$$

Now consider the rows $R_{m+1}, R_{m+2}, \ldots, R_{m+p+1}$ where $m \geq 0$. From the structure of H_f it is clear that these rows are obtained from $R_1, R_2, \ldots, R_{p+1}$ by omitting the first m columns. Thus for any $m \geq 0$,

$$R_{m+p+1} = \sum_{q=0}^{p-1} a_q R_{m+p-q} \tag{3.12}$$

That is, every row of H_f beginning with the (p+1)-st can be expressed as the linear combination of the preceding p rows given in (3.12). Therefore, each row beginning with the (p+1)-st can be expressed as a linear combination of the first p rows. But this implies that $p = n$. Writing (3.12) in terms of the elements of H_f gives (3.10), and it is clear that n is the least integer for which this can hold.

Now suppose (3.10) holds. Then every row of H_f can be expressed as a linear combination of the preceding n rows and thus of the first n rows. Thus rank $H_f \leq n$. But if rank $H_f < n$ then an expression of the form (3.10) would hold for an integer $< n$ so we must have rank $H_f = n$.

<u>Corollary</u> For any discrete-time signal f(k), rank $H_f = n$ iff the first n rows (columns) are linearly independent and row (column) n+1 can be expressed as a linear combination of them.

<u>Theorem 1</u> For a discrete-time signal f(k), rank $H_f = n < \infty$ iff the generating function of f(k), $F(z) = Z[f(k)]$ can be written as a unique reduced proper rational function of degree n.

<u>Proof</u>: Suppose f(k) has a reduced proper rational z-transform of degree n,

$$F(z) = \frac{b_n z^n + b_{n-1} z^{n-1} + \ldots + b_0}{z^n + a_{n-1} z^{n-1} + \ldots + a_1 z + a_0} = f(0) + f(1) z^{-1} + f(2) z^{-2} + \ldots \tag{3.13}$$

Then

$$b_n z^n + b_{n-1} z^{n-1} + \dots + b_0 = [z^n + a_{n-1} z^{n-1} + \dots + a_0][f(0) + f(1)z^{-1} + f(2)z^{-2} + \dots] \tag{3.14}$$

Equating coefficients of like powers of z gives the sets of equations

$$\begin{aligned} b_n &= f(0) \\ b_{n-1} &= a_{n-1}f(0) + f(1) \\ b_{n-2} &= a_{n-2}f(0) + a_{n-1}f(1) + f(2) \\ &\vdots \\ b_0 &= a_0 f(0) + a_1 f(1) + \dots + a_{n-1}f(n-1) + f(n) \end{aligned} \tag{3.15}$$

and

$$0 = a_0 f(k) + a_1 f(k+1) + \dots + a_{n-1} f(k+n-1) + f(k+n), \quad k = 1,2,3,\dots \tag{3.16}$$

Since F(z) is reduced, n is the least integer for which (3.16) can hold. Thus (3.16) satisfies the hypothesis of Lemma 1 and we have that rank $H_f = n$.

Now suppose H_f has finite rank n. Then, by Lemma 1 we can find $a_0, a_1, \dots, a_{n-1}$ such that an expression of the form (3.16) holds. We can define $b_0, b_1, \dots, b_n$ using (3.15). Then we have that

$$F(z) = \frac{b_n z^n + b_{n-1} z^{n-1} + \dots + b_0}{z^n + a_{n-1} z^{n-1} + \dots + a_0} = f(0) + f(1)z^{-1} + f(2)z^{-2} + \dots \tag{3.17}$$

and F(z) is reduced since n is the least integer for which (3.16) can hold. That F(z) is unique follows from the properties of division.

The proof of Theorem 1 gives a straightforward method for computing the z-transform of a given signal f(k). First n is determined by finding the rank of H_f. Then the n + 1 equations in (3.15) and n equations from (3.16) give 2n + 1 linear equations in the 2n + 1 coefficients of F(z). These equations are guaranteed to have a unique solution for the parameters $b_n, b_{n-1}, \dots, b_0, a_{n-1}, a_{n-2}, \dots, a_0$.

Example 1 For the unit pulse signal $u_0(k)$

$$\operatorname{rank} H_{u_0} = \operatorname{rank} \begin{bmatrix} 0 & 0 & \cdots \\ 0 & 0 & \cdots \\ \vdots & \vdots & \end{bmatrix} = 0$$

Thus n = 0 and from (3.15) we have that $b_0 = a_0 = 1$. Thus

$$U_0(z) = Z[u_0(k)] = 1$$

Example 2 The *unit step* signal is defined by

$$u_{-1}(k) = 1, \; k \geq 0$$

Thus

$$\operatorname{rank} H_{u_{-1}} = \operatorname{rank} \begin{bmatrix} 1 & 1 & 1 & \cdots \\ 1 & 1 & 1 & \cdots \\ 1 & 1 & 1 & \cdots \\ \vdots & \vdots & \vdots & \end{bmatrix} = 1$$

and from (3.15) and (3.16) with n = 1 we obtain

$$b_1 = f(0) = 1$$
$$b_0 = a_0 f(0) + f(1) = a_0 + 1$$
$$0 = a_0 f(1) + f(2) = a_0 + 1$$

This set of equations gives $b_1 = 1$, $b_0 = 0$, $a_0 = -1$ so that

$$U_{-1}(z) = Z[u_{-1}(k)] = \frac{z}{z - 1}$$

Example 3 Suppose λ is a real number and f(k) is the discrete exponential signal defined by

$$f(k) = \lambda^k, \quad k \geq 0$$

In this case, assuming $\lambda \neq 0$,

$$\operatorname{rank} H_f = \operatorname{rank} \begin{bmatrix} \lambda & \lambda^2 & \lambda^3 & \cdots \\ \lambda^2 & \lambda^3 & \lambda^4 & \cdots \\ \lambda^3 & \lambda^4 & \lambda^5 & \cdots \\ \vdots & \vdots & \vdots & \end{bmatrix} = 1$$

since the i-th row is simply λ^{i-1} times the first row.

Equations (3.15) and (3.16) with n = 1 readily yield

$$F(z) = \frac{z}{z - \lambda}$$

We now consider several operations on discrete-time signals using the z-transform representation. Thus we only consider signals that have finite rank Hankel matrices. It is left as an exercise to show directly that the signals resulting from these operations also have finite rank Hankel matrices. We will show this indirectly by manipulating the formal series representation of the original signal and its generating function to obtain a rational generating function for the formal series representation of the new signal.

Property 1 If $f(k)$ and $g(k)$ have proper rational z-transforms $F(z)$ and $G(z)$, then

$$\begin{aligned} Z[f(k) + g(k)] &= F(z) + G(z) \\ Z[af(k)] &= aF(z), \quad \text{for scalar } a \end{aligned} \tag{3.18}$$

Proof: Writing the formal series representation of $f(k) + g(k)$ as

$$\begin{aligned} f(k) + g(k) &= \sum_{i=0}^{\infty} [f(i) + g(i)]z^{-i} \\ &= \sum_{i=0}^{\infty} f(i)z^{-i} + \sum_{i=0}^{\infty} g(i)z^{-i} \end{aligned}$$

it is clear that the generating function is $F(z) + G(z)$ by the properties of long division. Similarly, writing

$$af(k) = \sum_{i=0}^{\infty} af(i)z^{-i} = a\sum_{i=0}^{\infty} f(i)z^{-i}$$

shows that the generating function for $af(k)$ is $aF(z)$.

Property 2 If $f(k)$ has a rational z-transform $F(z)$, then the z-transform of the unit delayed signal $f(k-1)$ is given by

$$Z[f(k-1)] = z^{-1}F(z) \tag{3.19}$$

Proof: We can write the unit delayed signal as

$$f(k-1) = \sum_{i=0}^{\infty} f(i-1)z^{-i} = z^{-1}\sum_{i=0}^{\infty} f(i-1)z^{-(i-1)}$$

$$= z^{-1}\sum_{i=1}^{\infty} f(i-1)z^{-(i-1)}$$

where we have used the one-sided signal convention to obtain the last equality. Changing the summation index to $j = i - 1$ gives

$$f(k-1) = z^{-1}\sum_{j=0}^{\infty} f(j)z^{-j}$$

Thus the generating function for $f(k-1)$ is $z^{-1}F(z)$ by the properties of long division.

Property 3 If $f(k)$ has a rational z-transform $F(z)$ then the unit advanced signal $f(k+1)$ has a rational z-transform given by

$$Z[f(k+1)] = zF(z) - zf(0) \tag{3.20}$$

Proof: We can write the unit advanced signal as

$$f(k+1) = \sum_{i=0}^{\infty} f(i+1)z^{-i}$$

Letting $j = i + 1$ be the index of summation gives

$$f(k+1) = \sum_{j=1}^{\infty} f(j)z^{-(j-1)}$$

$$= z\sum_{j=1}^{\infty} f(j)z^{-j}$$

$$= z\sum_{j=0}^{\infty} f(j)z^{-j} - zf(0)$$

Thus by the properties of long division we have

$$Z[f(k+1)] = zF(z) - zf(0)$$

Property 4 If f(k) and g(k) have rational z-transforms F(z) and G(z), and if

$$h(k) = \sum_{j=0}^{k} f(k-j)g(j), \quad k \geq 0 \tag{3.21}$$

then

$$Z[h(k)] = H(z) = F(z)G(z) \tag{3.22}$$

Proof: We can write the formal series representation of h(k) as

$$h(k) = \sum_{i=0}^{\infty} \sum_{j=0}^{i} f(i-j)g(j)z^{-i}$$

and using the fact that f(k) is a one-sided signal,

$$h(k) = \sum_{i=0}^{\infty} \sum_{j=0}^{\infty} f(i-j)g(j)z^{-i}$$

Interchanging the summation signs and changing the index i to q = i - j gives

$$h(k) = \sum_{j=0}^{\infty} \sum_{q=-j}^{\infty} f(q)g(j)z^{-(q+j)}$$

$$= \sum_{j=0}^{\infty} g(j)z^{-j} \sum_{q=-j}^{\infty} f(q)z^{-q}$$

Again by the one-sided convention we have

$$h(k) = \sum_{j=0}^{\infty} g(j)z^{-j} \sum_{q=0}^{\infty} f(q)z^{-q}$$

Thus, by the properties of long division we have

$$H(z) = G(z)F(z) = F(z)G(z)$$

From these properties it is clear that common operations on discrete-time signals have simple interpretations in terms of the z-transform representation. The remaining problem is that of finding the signal corresponding to a given rational z-transform, that is, the *inverse* z-transform.

Based on the definition of the z-transform, an obvious

method is to divide out the rational function. This gives the formal series representation from which the infinite sequence can be written. However, we will be more often interested in obtaining a closed form expression for the signal.

To do this we use the method of partial fraction expansion to express a z-transform as a sum of simple terms. If we can obtain a closed form expression for the signal corresponding to each term, then by Property 1 the sum of these is a closed form expression for the inverse z-transform. This procedure is not difficult since the kinds of terms that appear in the partial fraction expansion of a rational function are few in number and their inverse z-transforms are easily tabulated.

As indicated by the examples, most simple z-transforms have a factor of z in the numerator. Thus it is convenient to perform the partial fraction expansion on $z^{-1}F(z)$ and then multiply both sides by z so that a z appears in every numerator.

Example 4 Suppose we are given

$$F(z) = \frac{2z}{(z-2)(z-1)^2}$$

Using partial fraction expansion of $z^{-1}F(z)$ gives

$$z^{-1}F(z) = \frac{2}{(z-2)(z-1)^2}$$

$$= \frac{2}{z-2} - \frac{2}{z-1} - \frac{2}{(z-1)^2}$$

or

$$F(z) = \frac{2z}{z-2} - \frac{2z}{z-1} - \frac{2z}{(z-1)^2}$$

From Examples 2 and 3 we have the signals corresponding to the first two terms. It is readily verified (by division or by computation of the z-transform) that $z/(z-1)^2$ corresponds to the *unit ramp* signal $u_{-2}(k)$ defined by

$$u_{-2}(k) = k, \quad k \geq 0$$

Thus

$$f(k) = Z^{-1}[F(z)] = 2^{k+1} - 2 - 2k, \quad k \geq 0$$

The z-transform representation also can be used for signals with complex values. Although for our purposes these are not of interest in themselves, we often express a real valued signal as a sum of complex signals for simplicity. It is readily verified that using the rules of complex arithmetic, our development of the z-transform is valid for complex signals. Example 3 with λ complex serves as an illustration.

Example 5 Suppose we are given

$$F(z) = \frac{z}{z^2 - 2z + 2}$$

Then by partial fraction expansion of $z^{-1}F(z)$ we obtain

$$F(z) = -\frac{1}{2i}\frac{z}{z-(1-i)} + \frac{1}{2i}\frac{z}{z-(1+i)}$$

$$= -\frac{1}{2i}\frac{z}{z-\sqrt{2}\,e^{-i\pi/4}} + \frac{1}{2i}\frac{z}{z-\sqrt{2}\,e^{i\pi/4}}$$

Thus

$$f(k) = -\frac{1}{2i}(\sqrt{2}\,e^{-i\pi/4})^k + \frac{1}{2i}(\sqrt{2}\,e^{i\pi/4})^k$$

$$= \sqrt{2}^k \sin\left(\frac{k\pi}{4}\right), \quad k = 0, 1, \ldots$$

It should be noted that the z-transform of a vector signal is defined as the vector of z-transforms of the component signals. With this definition, the properties of the z-transform carry over to the vector/matrix notation. This can always be verified by writing the vector/matrix expression in scalar notation, taking the z-transform of each scalar expression, and rewriting the result in vector/matrix form.

3.3. STATE VECTOR EQUATIONS AND Z-TRANSFORMS

The state vector equation

$$\begin{aligned} x(k+1) &= Ax(k) + bu(k), \quad k = 0,1,2,\ldots \\ y(k) &= cx(k) + du(k), \quad x(0) = x_o \end{aligned} \tag{3.23}$$

may be viewed as a relationship between the vector signal x(k) and the scalar signals y(k) and u(k). We can express this relationship using the z-transform representation as follows. Let

$$X(z) = Z[x(k)] = \begin{bmatrix} Z[x_1(k)] \\ Z[x_2(k)] \\ \vdots \\ Z[x_n(k)] \end{bmatrix} \tag{3.24}$$

$$U(z) = Z[u(k)]$$

$$Y(z) = Z[y(k)]$$

Then from (3.23) we can write

$$zX(z) - zx_o = AX(z) + bU(z) \tag{3.25}$$

or

$$X(z) = z(zI - A)^{-1}x_o + (zI - A)^{-1}bU(z) \tag{3.26}$$

It will be shown below that $(zI - A)^{-1}$ is a matrix of strictly proper rational functions. Thus if U(z) is a proper rational function, X(z) is a vector of proper rational functions. Similarly

$$\begin{aligned} Y(z) &= cX(z) + dU(z) \\ &= cz(zI - A)^{-1}x_o + [c(zI - A)^{-1}b + d]U(z) \end{aligned} \tag{3.27}$$

is a proper rational function if U(z) is a proper rational function. The expression in (3.27) can be used to compute the response of a system in closed form in simple situations.

<u>Example 6</u> Consider the state vector equation (RCF)

$$x(k+1) = \begin{bmatrix} 0 & 1 \\ -12 & 7 \end{bmatrix} x(k) + \begin{bmatrix} 0 \\ 1 \end{bmatrix} u(k)$$

$$y(k) = \begin{bmatrix} 1 & 0 \end{bmatrix} x(k)$$

with $x(0) = 0$ and $u(k) = 1$ for all $k \geq 0$. In this case (3.27) simplifies to

$$Y(z) = c(zI - A)^{-1}bU(z)$$

where $U(z) = \frac{z}{z - 1}$ from Example 2. An easy computation gives that

$$(zI - A)^{-1} = \begin{bmatrix} \frac{z - 7}{z^2 - 7z + 12} & \frac{1}{z^2 - 7z + 12} \\ \frac{-12}{z^2 - 7z + 12} & \frac{z}{z^2 - 7z + 12} \end{bmatrix}$$

so that

$$Y(z) = \frac{z}{(z^2 - 7z + 12)(z - 1)}$$

$$= \frac{1}{3}\frac{z}{z - 4} - \frac{1}{2}\frac{z}{z - 3} + \frac{1}{6}\frac{z}{z - 1}$$

Thus the response is given by

$$y(k) = \frac{1}{3}(4)^k - \frac{1}{2}(3)^k + \frac{1}{6}, \quad k \geq 0$$

By comparing the solution formula derived in Chap. 1 with (3.26) when $u(k) = 0$ for all k and x_o is arbitrary,

$$\begin{aligned} x(k) &= A^k x_o \\ X(z) &= z(zI - A)^{-1}x_o \end{aligned} \tag{3.28}$$

we conclude that the z-transform of the matrix signal A^k is

$$Z[A^k] = z(zI - A)^{-1} \tag{3.29}$$

(This can be shown directly; see Problem 11.) Using (3.29) we can obtain a closed form expression for A^k. Let $d(z)$ be the characteristic polynomial of A,

$$\begin{aligned} d(z) &= \det (zI - A) \\ &= \prod_{i=1}^{m} (z - \lambda_i)^{\sigma_i}, \quad \text{degree } d(z) = n \end{aligned} \tag{3.30}$$

where each λ_i is a distinct eigenvalue of A with multiplicity σ_i. Let B(z) be the classical adjoint of (zI - A), each element $b_{kq}(z)$ of which is a polynomial in z of degree n - 1 or less. Then

$$(zI - A)^{-1} = \frac{B(z)}{d(z)} \tag{3.31}$$

is a matrix of strictly proper rational functions. Thus we can expand each element $b_{kq}(z)/d(z)$ in (3.31) in a partial fraction expansion to obtain, after multiplying both sides by z,

$$\begin{aligned} z\frac{b_{kq}(z)}{d(z)} &= \frac{zw_{kq}^{11}}{z - \lambda_1} + \frac{zw_{kq}^{12}}{(z - \lambda_1)^2} + \cdots + \frac{zw_{kq}^{1\sigma_1}}{(z - \lambda_1)^{\sigma_1}} \\ &+ \cdots + \frac{zw_{kq}^{m1}}{z - \lambda_m} + \frac{zw_{kq}^{m2}}{(z - \lambda_m)^2} + \cdots + \frac{zw_{kq}^{m\sigma_m}}{(z - \lambda_m)^{\sigma_m}} \end{aligned}$$

Performing this expansion for each k, q = 1, ..., n we define the following matrices of the partial fraction expansion coefficients. Let W_{11} be the n × n matrix with k, q entry w_{kq}^{11}, let W_{12} be the n × n matrix with k, q entry w_{kq}^{12}, and so on. Using this matrix notation we can arrange all the partial fraction expansions into the equation

$$z(zI - A)^{-1} = z\frac{B(z)}{d(z)} = \sum_{i=1}^{m} \sum_{j=1}^{\sigma_i} W_{ij} \frac{z}{(z - \lambda_i)^j} \tag{3.32}$$

Now we can take the inverse z-transforms of both sides of (3.32) to obtain an expression for A^k. If each $\sigma_i = 1$ then m = n and (3.32) gives

$$A^k = \sum_{i=1}^{n} W_{i1} (\lambda_i)^k, \quad k \geq 0 \tag{3.33}$$

This expression should be compared with that derived in

Chap. 1 under the same assumption of distinct eigenvalues. For the case of multiplicities greater than 1, (3.32) readily yields an expression for A^k upon determination of the inverse z-transform of $z/(z - \lambda_i)^j, j \geq 2$. (See Problems 6,9)

3.4. THE TRANSFER FUNCTION

We now return to the problem of formulating input/output maps corresponding to a state vector equation.

Definition 3 The *transfer function* H(z) corresponding to a discrete-time state vector equation (A,b,c,d) is the quotient Y(z)/U(z) under the assumption that $x_o = 0$. That is,

$$H(z) = \frac{Y(z)}{U(z)} = c(zI - A)^{-1}b + d \tag{3.34}$$

Thus the transfer function corresponding to a given state vector equation is a proper (strictly proper if d = 0) rational function of the indeterminate z. From the solution expression in (3.27), it is clear that H(z) completely characterizes the input/output behavior for all inputs with proper rational z-transforms. It should be noted that with $x_o = 0$ and $u(k) = u_0(k)$, the unit pulse input, we have Y(z) = H(z) since $U_0(z) = 1$. That is, the transfer function is precisely the z-transform of the unit pulse response.

The following terminology is very common in discussing a transfer function H(z). The roots of the numerator polynomial of the reduced form of H(z) are called the *zeros* of H(z). The roots of the denominator polynomial of the reduced form of H(z) are called the *poles* of H(z).

It is intuitively clear that the input/output map corresponding to a state vector equation should not be altered by a change of state variables. This is readily demonstrated for the transfer function input/output map by comparing the transfer functions corresponding to the state vector equations $(P^{-1}AP, P^{-1}b, cP, d)$ and (A,b,c,d):

$$
\begin{aligned}
(cP)[zI - (P^{-1}AP)]^{-1}(P^{-1}b) + d &= (cP)[P^{-1}(zI - A)P]^{-1}(P^{-1}b) + d \\
&= (cP)[P^{-1}(zI - A)^{-1}P](P^{-1}b) + d \\
&= c(zI - A)^{-1}b + d
\end{aligned}
\tag{3.35}
$$

Example 7 Consider the fish hatchery model of Example 2, Chap. 2 with the simplifying assumptions that $a_1 = a_3 = 0$ and $a_6 = 1$ (the adult fish are sterile and live one year, and the young do not eat fry). In this case the state vector equation is

$$
x(k+1) = \begin{bmatrix} 0 & -a_2 & 0 \\ 1 & -a_4 & 0 \\ 0 & a_5 & 0 \end{bmatrix} x(k) + \begin{bmatrix} 1 \\ 0 \\ 0 \end{bmatrix} u(k)
$$

$$
y(k) = \begin{bmatrix} 0 & a_4 & 0 \end{bmatrix} x(k)
$$

A straightforward calculation gives

$$
(zI - A)^{-1} = \frac{\begin{bmatrix} z^2+a_4z & -a_2z & 0 \\ z & z^2 & 0 \\ a_5 & a_5z & z^2+a_4z+a_2 \end{bmatrix}}{z(z^2 + a_4z + a_2)}
$$

Thus the hatchery transfer function is

$$
H(z) = c(zI - A)^{-1}b = \frac{a_4z}{z(z^2 + a_4z + a_2)} = \frac{a_4}{z^2 + a_4z + a_2}
$$

Suppose there is no initial stocking $x_o = 0$, and $u(k) = u_0(k) = (1, 0, 0, \ldots)$. Then by dividing $H(z)$ it is found that the output of the hatchery is, in this case,

$$
y(k) = (0, 0, a_4, -a_4^2, \ldots)
$$

Under the assumptions made in the development of the model, a_4 is positive so in year $k = 3$ a negative number of young fish are produced. Of course this is impossible in reality. To see why the model breaks down in this situation, the reader should trace each sub-population for years 0, 1, 2, and 3. Note that each year enough eggs should be supplied to feed the fry in order for the model to be valid.

The RCF and OCF state vector equations discussed in Chap. 2 have particularly close connections with the transfer function input/output map. We first compute the transfer function of the general RCF canonical form. To find an expression for $(zI - A)^{-1}b$, consider the equation

$$(zI - A)v = b \tag{3.36}$$

where v is an $n \times 1$ vector. In scalar notation, (3.36) becomes

$$\begin{bmatrix} zv_1 \\ zv_2 \\ \vdots \\ zv_n \end{bmatrix} - \begin{bmatrix} 0 & 1 & 0 & \cdots & 0 \\ 0 & 0 & 1 & \cdots & 0 \\ \vdots & \vdots & \vdots & & \vdots \\ -a_0 & -a_1 & -a_2 & \cdots & -a_{n-1} \end{bmatrix} \begin{bmatrix} v_1 \\ v_2 \\ \vdots \\ v_n \end{bmatrix} = \begin{bmatrix} 0 \\ 0 \\ \vdots \\ 1 \end{bmatrix} \tag{3.37}$$

Thus

$$\begin{aligned} v_{i+1} &= zv_i, \quad i = 1, 2, \ldots, n-1 \\ zv_n + a_0v_1 + a_1v_2 &+ \ldots + a_{n-1}v_n = 1 \end{aligned} \tag{3.38}$$

which gives

$$z^nv_1 + a_0v_1 + a_1zv_1 + \ldots + a_{n-1}z^{n-1}v_1 = 1 \tag{3.39}$$

or

$$v_1 = \frac{1}{z^n + a_{n-1}z^{n-1} + \ldots + a_1z + a_0} \tag{3.40}$$

From (3.40) and (3.38) we have

$$v = (zI - A)^{-1}b = \frac{1}{z^n + a_{n-1}z^{n-1} + \ldots + a_0} \begin{bmatrix} 1 \\ z \\ \vdots \\ z^{n-1} \end{bmatrix}$$

and since $c = [c_0 \; c_1 \; \cdots \; c_{n-1}]$

$$c(zI - A)^{-1}b + d = \frac{c_{n-1}z^{n-1} + \dots + c_1 z + c_0}{z^n + a_{n-1}z^{n-1} + \dots + a_1 z + a_0} + d \tag{3.41}$$

We have shown that the transfer function of an RCF state vector equation can be written by inspection. A similar computation shows that the transfer function of the OCF state vector equation with

$$A = \begin{bmatrix} 0 & 0 & \dots & 0 & -a_0 \\ 1 & 0 & \dots & 0 & -a_1 \\ 0 & 1 & \dots & 0 & -a_2 \\ \vdots & \vdots & & \vdots & \vdots \\ 0 & 0 & \dots & 1 & -a_{n-1} \end{bmatrix} \quad b = \begin{bmatrix} c_0 \\ c_1 \\ \vdots \\ c_{n-1} \end{bmatrix} \quad c = [0\ 0\ \dots\ 1]$$

is precisely that in (3.41). Thus given a proper rational transfer function $H(z)$ we can divide to obtain $H(z) = G(z) + d$ where $G(z)$ is strictly proper. Then the corresponding RCF or OCF state vector equation can be written by inspection.

3.5. THE BLOCK DIAGRAM

Often a system is composed of several subsystems each of which is described by a transfer function. In this situation a *block diagram* is often used to illustrate the interconnections. A block diagram is composed of two types of elements: transfer function blocks and adders. The transfer function block is used to illustrate the relation

$$Y(z) = H(z)U(z)$$

and is usually drawn as in Fig. 3-1. As in the SVD, an adder

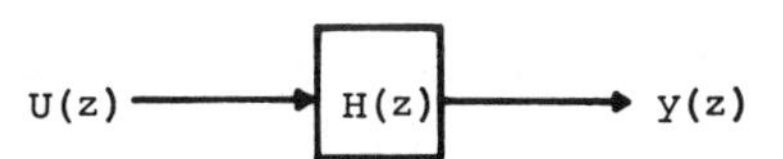

FIG. 3-1. A transfer function block.

indicates the signed addition of signals but in a block diagram the z-transform notation is used as indicated in Fig.3-2.

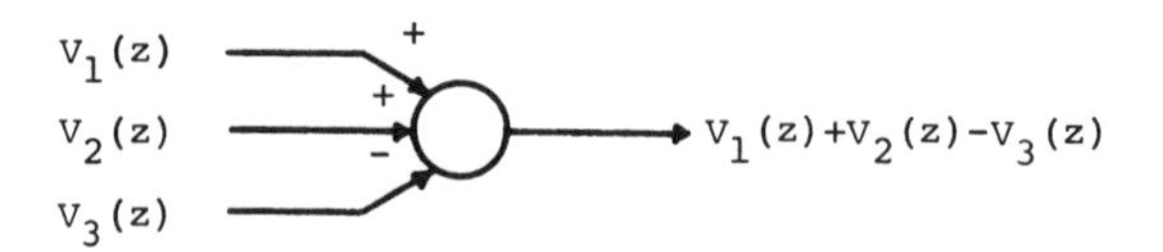

FIG. 3-2. An adder.

Using these definitions it is readily seen that the *cascade* connection shown in Fig. 3-3 represents the relation

$$Y(z) = H_2(z)H_1(z)U(z)$$

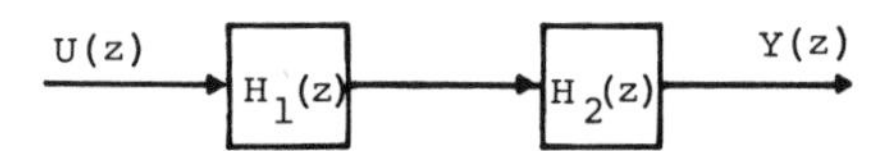

FIG. 3-3. Two transfer functions in cascade.

The *parallel* connection shown in Fig. 3-4 represents

$$Y(z) = [H_1(z) + H_2(z)]U(z)$$

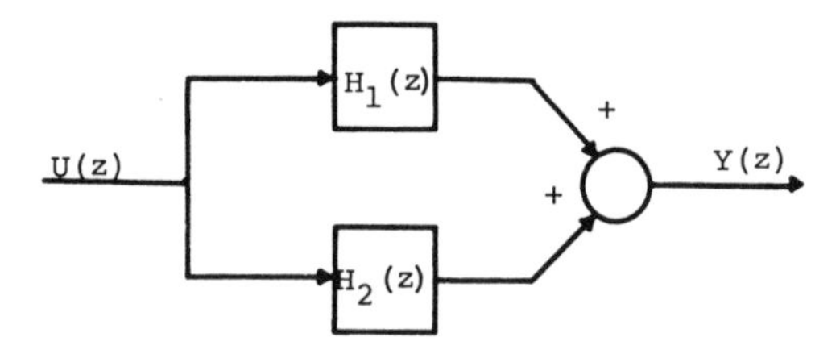

FIG. 3-4. Two transfer functions in parallel.

In more complex situations it is often helpful to label internal signals in the diagram when calculating the overall transfer function.

Example 8 Given the block diagram shown in Fig. 3-5, we can calculate the transfer function of the interconnection as follows:

$$Y(z) = H_1(z)E(z)$$

$$E(z) = U(z) - H_2(z)Y(z)$$

$$Y(z) = H_1(z)U(z) - H_1(z)H_2(z)Y(z)$$

$$\frac{Y(z)}{U(z)} = \frac{H_1(z)}{1 + H_1(z)H_2(z)}$$

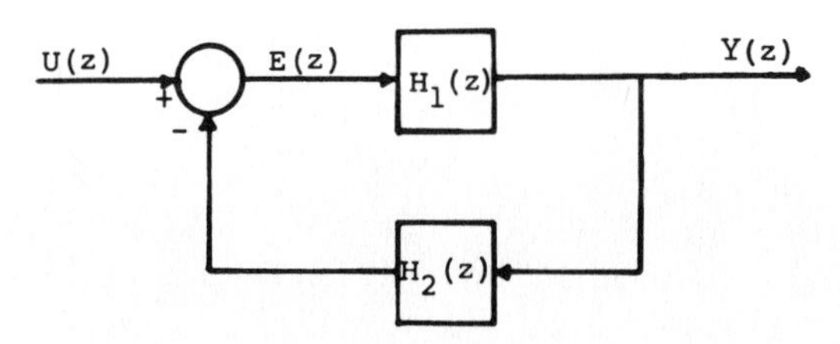

FIG. 3-5. A feedback block diagram.

Example 9 The block diagram shown in Fig. 3-6 is more complicated but can still be handled in easy steps using the labelled signals.

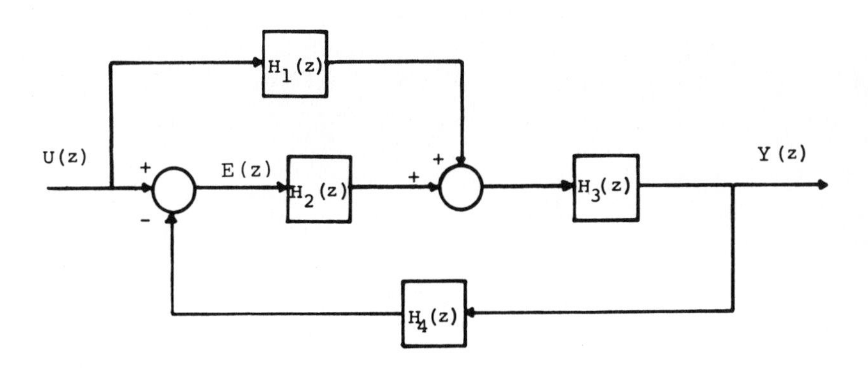

FIG. 3-6. Block diagram.

$$Y(z) = H_3(z)[H_1(z)U(z) + H_2(z)E(z)]$$

$$E(z) = U(z) - H_4(z)Y(z)$$

$$Y(z) = H_3(z)[H_1(z)U(z) + H_2(z)U(z) - H_2(z)H_4(z)Y(z)]$$

$$\frac{Y(z)}{U(z)} = \frac{[H_1(z) + H_2(z)]H_3(z)}{1 + H_2(z)H_3(z)H_4(z)}$$

3.6. THE UNIT IMPULSE RESPONSE

For the continuous-time state vector equation

$$\begin{aligned} \dot{x}(t) &= Ax(t) + bu(t) \quad t \geq 0 \\ y(t) &= cx(t) + du(t) \end{aligned} \tag{3.42}$$

the response to any piecewise continuous input when $x(0) = 0$ is given by

$$y(t) = \int_0^t h(t-\sigma)u(\sigma)d\sigma + du(t) \tag{3.43}$$

where $h(t) = ce^{At}b$. Thus if d and h(t), $t \geq 0$ are known, the response corresponding to any input can be obtained.

However, unlike the discrete-time case, considerable difficulties arise when an input which yields the output $y(t) = h(t)$ is sought. In fact, under our definition of a continuous-time signal, no such input signal exists. To rectify this situation in a rigorous manner requires a long digression into the theory of generalized functions and the impulse function. We will be content to indicate the nature of the situation.

Consider an input of the form of a rectangular pulse with unit area as depicted in Fig. 3-7. For $t > 1/\Delta$ the

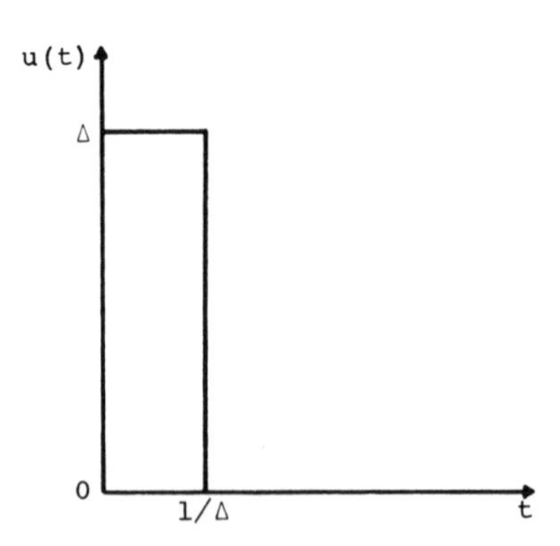

FIG. 3-7. Unit area pulse.

response of the system is

$$y(t) = \int_0^{1/\Delta} h(t-\sigma)\Delta \, d\sigma, \quad t > 1/\Delta$$

Letting Δ become very large, we can approximate this expression by

$$y(t) \simeq \frac{1}{\Delta} h(t)\Delta = h(t), \; t > 0$$

Thus the response to a very tall, very thin pulse of unit area is very nearly $h(t)$ for $t > 0$. It is difficult to analyze $y(t)$ at $t = 0$ if $d \neq 0$ because the second term in (3.43) increases directly with Δ. If $d = 0$, then we can state that the response to a tall thin pulse of unit area is approximately $y(t) = h(t)$, $t \geq 0$. Therefore we call $h(t)$ the unit impulse response.

3.7. THE S-TRANSFORM

The s-transform representation (see Table 3-2, at end of chapter) involves assigning to a continuous-time signal $f(t)$ a unique function $F(s)$ of the indeterminate s. As in the discrete-time case, the s-transform $F(s)$ is obtained as a generating function for a formal series. Common operations on continuous-time signals are then shown to correspond to simple operations on the s-transforms of the signals.

We consider only one-sided continuous-time signals which are analytic functions for $0 \leq t < \infty$. That is, they are infinitely differentiable for $0 \leq t < \infty$ and they have convergent Taylor's series in some neighborhood about any t_0, $0 \leq t_0 < \infty$. It is well known that such functions are completely determined by their Taylor's series expansion about $t = 0$,

$$f(t) = f(0) + f^{(1)}(0)\frac{t}{1!} + f^{(2)}(0)\frac{t^2}{2!} + \ldots \qquad (3.44)$$

It is understood that the derivatives are all defined from the right since $f(t) = 0$ for $t < 0$.

We will also represent $f(t)$ as a formal power series involving a place marker to indicate the correspondence with

(3.44). Let s^{-k-1} denote the k-th term in (3.44), $k \geq 0$. That is, replace $t^k/k!$ by $s^{-(k+1)}$. Then we write the formal series

$$f(t) = \sum_{i=0}^{\infty} f^{(i)}(0)s^{-(i+1)} \tag{3.45}$$

where

$$f^{(0)}(0) = f(0) \tag{3.46}$$

As in the z-transform case, the summation sign in (3.45) indicates a formal series and not a numerical total. Thus convergence considerations will not arise.

The advantage of the formal series representation is that under certain conditions we can find a rational generating function F(s) which when divided out yields the series. In this event we call F(s) the s-transform of f(t) and write F(s) = S[f(t)].

Definition 4 The *Hankel matrix* H_f associated with an analytic continuous-time signal f(t) is the symmetric matrix with an infinite number of rows and columns given by

$$H_f = \begin{bmatrix} f(0) & f^{(1)}(0) & f^{(2)}(0) & \cdots \\ f^{(1)}(0) & f^{(2)}(0) & f^{(3)}(0) & \cdots \\ f^{(2)}(0) & f^{(3)}(0) & f^{(4)}(0) & \cdots \\ \vdots & \vdots & \vdots & \end{bmatrix} \tag{3.47}$$

A proper rational function F(s) is defined precisely as in Sec. 3.2 and written in the general form

$$F(s) = \frac{b_n s^n + b_{n-1}s^{n-1} + \ldots + b_1 s + b_0}{s^n + a_{n-1}s^{n-1} + \ldots + a_1 s + a_0} \tag{3.48}$$

The key mathematical results from Sec. 3.2 clearly apply here after a change in notation. We will simply restate the results and leave the proofs to the reader.

Lemma 2 For any analytic continuous-time signal f(t), H_f

has finite rank n iff there exist n scalars $a_0, a_1, \ldots, a_{n-1}$ such that

$$f^{(k)}(0) = \sum_{q=0}^{n-1} a_q f^{(k-q-1)}(0), \quad k = n, n+1, \ldots \tag{3.49}$$

and n is the least number having this property.

Corollary For any analytic continuous-time signal f(t), rank $H_f = n$ iff the first n rows (columns) of H_f are linearly independent and row (column) n + 1 can be expressed as a linear combination of them.

Theorem 2 For an analytic continuous-time signal f(t), rank $H_f = n < \infty$, iff the generating function of f(t), F(s) = S[f(t)] can be written as a unique reduced proper rational function of degree n.

Given an analytic continuous-time signal f(t) with rank $H_f = n$, F(s) is readily computed from 2n + 1 equations derived from

$$F(s) = \frac{b_n s^n + \ldots + b_1 s + b_0}{s^n + a_{n-1}s^{n-1} + \ldots + a_1 s + a_0} = f(0)s^{-1} + f^{(1)}(0)s^{-2} + f^{(2)}(0)s^{-3} + \ldots \tag{3.50}$$

For reference, these equations are listed below:

$$\begin{aligned}
b_n &= 0 \\
b_{n-1} &= f(0) \\
b_{n-2} &= a_{n-1}f(0) + f^{(1)}(0) \\
b_{n-3} &= a_{n-2}f(0) + a_{n-1}f^{(1)}(0) + f^{(2)}(0) \\
&\vdots \\
b_0 &= a_1 f(0) + a_2 f^{(1)}(0) + \ldots + a_{n-1}f^{(n-2)}(0) + f^{(n-1)}(0)
\end{aligned} \tag{3.51}$$

and

$$0 = a_0 f^{(k)}(0) + a_1 f^{(k+1)}(0) + \ldots + a_{n-1} f^{(k+n-1)}(0) + f^{(k+n)}(0)$$

$$k = 0, 1, 2, \ldots \tag{3.52}$$

Note that since $b_n = 0$, the s-transform of an analytic continuous-time signal is always a strictly proper rational function.

Example 10 The *unit step* signal is defined by

$$u_{-1}(t) = 1, \quad t \geq 0$$

Clearly

$$\text{rank } H_{u_{-1}} = \text{rank} \begin{bmatrix} 1 & 0 & 0 & \cdots \\ 0 & 0 & 0 & \cdots \\ 0 & 0 & 0 & \cdots \\ \vdots & \vdots & \vdots & \end{bmatrix} = 1$$

From (3.51) and (3.52) with n = 1 we have the equations

$$b_1 = 0$$

$$b_0 = 1$$

$$0 = a_0$$

Thus $U_{-1}(s) = S[u_{-1}(t)] = \frac{1}{s}$.

Example 11 For $f(t) = e^{\lambda t}$,

$$\text{rank } H_f = \text{rank} \begin{bmatrix} 1 & \lambda & \lambda^2 & \cdots \\ \lambda & \lambda^2 & \lambda^3 & \cdots \\ \lambda^2 & \lambda^3 & \lambda^4 & \cdots \\ \vdots & \vdots & \vdots & \end{bmatrix} = 1$$

since each row can be expressed as a scalar multiple of the first row. From (3.51) and (3.52) we obtain the equations

$$b_1 = 0$$

$$b_0 = 1$$

$$0 = a_0 + \lambda$$

Thus

$$F(s) = \frac{1}{s - \lambda}$$

We now consider several common operations on analytic continuous-time signals using the s-transform representation. We only consider signals with finite rank Hankel matrices. Since the proofs proceed in a manner similar to the z-transform case, several are left as exercises.

Property 5 If f(t) and g(t) have strictly proper rational s-transforms F(s) and G(s), then

$$\begin{aligned} S[f(t) + g(t)] &= F(s) + G(s) \\ S[af(t)] &= aF(s), \quad \text{for scalar } a \end{aligned} \tag{3.53}$$

Proof: See Problem 16.

Property 6 If f(t) has a strictly proper rational s-transform F(s), and if

$$g(t) = \int_0^t f(\sigma)\,d\sigma \tag{3.54}$$

then g(t) has a strictly proper rational s-transform given by

$$G(s) = s^{-1}F(s) \tag{3.55}$$

Proof: Since

$$f(t) = f(0) + f^{(1)}(0)\,\frac{t}{1!} + f^{(2)}(0)\,\frac{t^2}{2!} + \ldots$$

we have

$$g(t) = f(0)\,\frac{t}{1!} + f^{(1)}(0)\,\frac{t^2}{2!} + f^{(2)}(0)\,\frac{t^3}{3!} + \ldots$$

The formal series representation of g(t) is

$$g(t) = \sum_{i=0}^{\infty} f^{(i)}(0)s^{-(i+2)} = s^{-1}\sum_{i=0}^{\infty} f^{(i)}(0)s^{-(i+1)}$$

By the properties of long division, it is clear that

$$S[g(t)] = s^{-1}F(s)$$

Property 7 If f(t) has a strictly proper rational s-transform F(s), then

$$S[\frac{d}{dt}f(t)] = sF(s) - f(0) \tag{3.56}$$

Proof: See Problem 17.

Property 8 If f(t) and g(t) have strictly proper rational s-transforms F(s) and G(s), and if

$$h(t) = \int_0^t f(t-\sigma)g(\sigma)d\sigma \tag{3.57}$$

then

$$S[h(t)] = F(s)G(s) \tag{3.58}$$

Proof: See Problem 18.

The problem of obtaining the inverse s-transform, that is, the analytic continuous-time signal corresponding to a given strictly proper rational F(s), is handled much like the inverse z-transform. However in this case simple division is not as useful since it yields only the values f(0), $f^{(1)}(0)$, ... It is only possible to recognize a few types of signals from their Taylor's expansions. Fortunately, using partial fraction expansion, only a few kinds of signals need be known.

The s-transform can be used for complex valued signals. This follows from the basic mathematical development, and Example 10 with complex λ serves as an illustration. As in the discrete-time case, complex signals are sometimes useful in the representation of real valued signals.

Example 12 Suppose we are given

$$F(s) = \frac{s}{s^2 - 2s + 2}$$

Then by partial fraction expansion,

$$\frac{s}{s^2 - 2s + 2} = \frac{s}{[s - (1+i)][s - (1-i)]} = \frac{1/2 - i(1/2)}{s - (1+i)} + \frac{1/2 + i(1/2)}{s - (1-i)}$$

which gives

$$f(t) = \left(\frac{1}{2} - i\,\frac{1}{2}\right)e^{(1+i)t} + \left(\frac{1}{2} + i\,\frac{1}{2}\right)e^{(1-i)t}$$

$$= e^{t}\left(\frac{e^{it} + e^{-it}}{2} + \frac{e^{it} - e^{-it}}{2i}\right)$$

$$= e^{t}\cos(t) + e^{t}\sin(t)$$

Finally, we note that the s-transform of a vector signal is defined as the vector of s-transforms of the component signals.

3.8. STATE VECTOR EQUATIONS AND S-TRANSFORMS

The state vector equation

$$\begin{aligned} \dot{x}(t) &= Ax(t) + bu(t), \quad t \geq 0 \\ y(t) &= cx(t) + du(t), \quad x(0) = x_o \end{aligned} \tag{3.59}$$

may be viewed as a relationship between the signals $x(t)$, $u(t)$, and $y(t)$. Letting

$$X(s) = S[x(t)] = \begin{bmatrix} S[x_1(t)] \\ S[x_2(t)] \\ \vdots \\ S[x_n(t)] \end{bmatrix} \tag{3.60}$$

$$U(s) = S[u(t)]$$

$$Y(s) = S[y(t)]$$

we can write (3.59) as

$$\begin{aligned} sX(s) - x_o &= AX(s) + bU(s) \\ Y(s) &= cX(s) + dU(s) \end{aligned} \tag{3.61}$$

assuming that all the signals have strictly proper rational s-transforms. Manipulating (3.61) gives

$$X(s) = (sI - A)^{-1}x_o + (sI - A)^{-1}bU(s) \tag{3.62}$$

and

$$Y(s) = c(sI - A)^{-1}x_o + [c(sI - A)^{-1}b + d]U(s) \tag{3.63}$$

Since $(sI - A)^{-1}$ is a matrix of strictly proper rational functions, it is clear from (3.62) and (3.63) that if $U(s)$ is strictly proper then $Y(s)$ and the components of $X(s)$ are strictly proper rational functions.

We can use the s-transform to obtain a closed form expression for the solution of a state vector equation in simple situations. Also, we can obtain a closed form expression for the matrix exponential e^{At}. By a direct computation (see Problem 21) or by comparing (3.62) with the solution formulas in Chap. 1, we have

$$S[e^{At}] = (sI - A)^{-1} \tag{3.64}$$

Writing

$$(sI - A)^{-1} = \frac{B(s)}{d(s)} \tag{3.65}$$

where $d(s) = \det(sI - A) = \prod_{i=1}^{m} (s - \lambda_i)^{\sigma_i}$, and $B(s)$ is the classical adjoint of $(sI - A)$, we find by partial fraction expansion that

$$(sI - A)^{-1} = \sum_{i=1}^{m} \sum_{j=1}^{\sigma_i} W_{ij} \frac{1}{(s - \lambda_i)^j} \tag{3.66}$$

where each W_{ij} is an $n \times n$ matrix of partial fraction expansion coefficients. Taking the inverse s-transform of (3.66) gives

$$e^{At} = \sum_{i=1}^{m} \sum_{j=1}^{\sigma_i} W_{ij} \frac{t^{j-1}}{(j-1)!} e^{\lambda_i t} \tag{3.67}$$

It is interesting to compare (3.67) with the closed form expression derived in Chap. 1 under the assumption of distinct eigenvalues.

3.9. TRANSFER FUNCTIONS AND BLOCK DIAGRAMS

Definition 5 The *transfer function* H(s) corresponding to a continuous-time state vector equation (A,b,c,d) is the quotient Y(s)/U(s) under the assumption that $x_o = 0$. That is

$$H(s) = \frac{Y(s)}{U(s)} = c(sI - A)^{-1}b + d \tag{3.68}$$

Thus for all input signals with strictly proper rational s-transforms, the transfer function completely characterizes the input/output behavior. Since (3.68) has precisely the same form as the discrete-time transfer function, the calculations in Sec. 3.4 need not be repeated. In particular, two state vector equations related by a change of variables have precisely the same transfer functions, and the transfer functions of the RCF and OCF state vector equations can be written by inspection. Also the notions of block diagrams and block diagram calculations are identical to those in the discrete-time case. We simply replace every z in Sec. 3.5 by s. The poles and zeros of H(s) are defined as in the discrete-time case.

However there is one facet of the continuous-time case which should be noted. Every continuous-time signal with a finite rank Hankel matrix has a strictly proper s-transform. If a transfer function H(s) is strictly proper, then from Property 8 it is clear that H(s) = S[h(t)], where h(t) is the unit impulse response discussed in Sec. 3.6. If H(s) has equal degree numerator and denominator, then it cannot correspond to the s-transform of a continuous-time signal under our definitions. However in this case it can be shown that H(s) = S[h(t)] + d. Recall that in the discrete-time case this distinction is not necessary since H(z) = Z[h(k)] for nonzero d as well as d = 0.

Example 13 Suppose we wish to compute the transfer function of the bucket system shown in Figure 3-8. Rather than write the four-dimensional state vector equation description and compute $c(sI - A)^{-1}b + d$, we note that this is a cascade

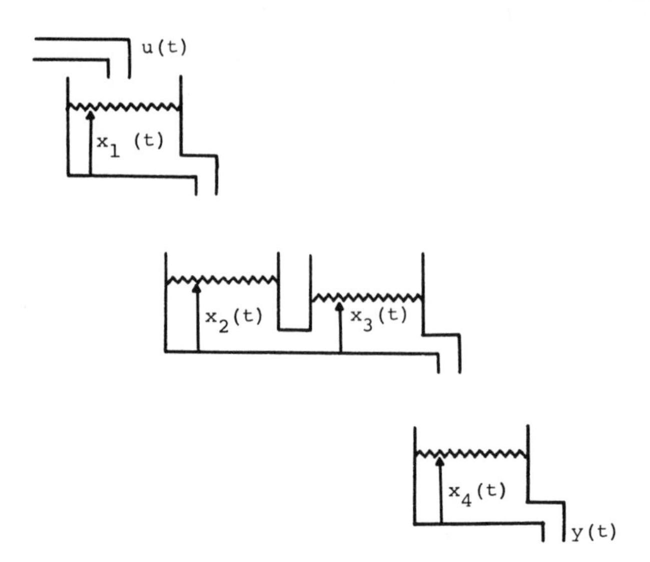

FIG. 3-8. Bucket system with all parameters unity.

connection of three simple subsystems. Thus we can compute the transfer function of each subsystem and multiply to obtain the overall system transfer function description.

From Example 6, Chap. 1, each single bucket subsystem is described by (A,b,c,d) = (-1,1,1,0) and thus has the transfer function

$$c(sI - A)^{-1}b + d = \frac{1}{s + 1}$$

Also from this example, the parallel bucket subsystem is described by

$$(A,b,c,d) = \begin{bmatrix} -1 & 1 \\ 1 & -2 \end{bmatrix}, \begin{bmatrix} 1 \\ 0 \end{bmatrix}, \begin{bmatrix} 0 & 1 \end{bmatrix}, 0$$

Thus the transfer function description of this subsystem is

$$c(sI - A)^{-1}b + d = \begin{bmatrix} 0 & 1 \end{bmatrix} \begin{bmatrix} s + 1 & -1 \\ -1 & s + 2 \end{bmatrix}^{-1} \begin{bmatrix} 1 \\ 0 \end{bmatrix} = \frac{1}{s^2 + 3s + 1}$$

and the overall system transfer function description is

$$\frac{Y(s)}{U(s)} = H(s) = \frac{1}{(s + 1)^2(s^2 + 3s + 1)} = \frac{1}{s^4 + 5s^3 + 8s^2 + 5s + 1}$$

Note that this transfer function and thus the system input/output behavior is independent of the ordering of the subsystems in the cascade connection.

PROBLEMS

1. Find the response of a discrete-time state vector equation to a unit delayed unit pulse input.
2. Show that two discrete-time state vector equations related by a nonsingular change of variables have the same unit pulse response.
3. Find the z-transform of a signal of *finite length*,

   $f(k) = (f(0), f(1), \ldots, f(n), 0, 0, \ldots)$
4. Find the z-transform of the *unit ramp* signal defined by $u_{-2}(k) = k, \quad k \geq 0$.
5. Find the z-transform of the signal $f(k) = (1, 0, \lambda, 0, \lambda^2, 0, \lambda^3, 0, \ldots)$.
6. Find the z-transform of the signal defined by $f(k) = k\lambda^{k-1}, \; k \geq 0$.
7. Show that if f(k) has a proper rational z-transform F(z), then $Z[a^k f(k)] = F(a^{-1}z)$, for scalar a.
8. If f(k) has a rational z-transform F(z), and g(k) is defined by $g(k) = \sum_{j=0}^{k} f(j)$, find $G(z) = Z[g(k)]$.
9. Find a general form for the inverse z-transform of $F(z) = z/(z - \lambda)^j, \; j \geq 2$.
10. Find a closed form expression for the inverse z-transform of $F(z) = 2z(z^2 - 1)/(z^2 + 1)^2$.
11. Show that $z(zI - A)^{-1}$ is the generating function for the formal series representation of the matrix signal $f(k) = A^k$.
12. Find the transfer function Y(z)/U(z) for the block diagram shown in Fig. 3-9.

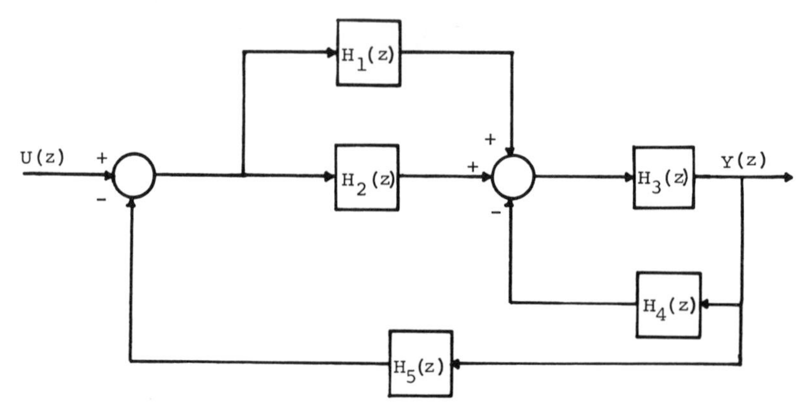

FIG. 3-9. Block diagram for Problem 12.

13. Find the s-transform of the *unit ramp* signal defined by $f(t) = t,\ t \geq 0$.

14. Find the s-transform of the signal $f(t) = te^{\lambda t},\ t \geq 0$.

15. Find the s-transform of the polynomial signal $f(t) = a_0 + a_1 t + \ldots + a_n t^n,\ t \geq 0$.

16. Prove Property 5.

17. Prove Property 7.

18. Prove Property 8.

19. Show that if f(t) has a strictly proper rational s-transform F(s), then $S[e^{-at}f(t)] = F(s+a)$, for scalar a.

20. Find the inverse s-transform of $F(s) = s/(s^2 + a^2)$ for scalar a.

21. Using the formal series representation of e^{At}, show that $S[e^{At}] = (sI - A)^{-1}$.

22. Rework Example 6 assuming that the state vector equation is continuous-time and $u(t) = 1$ for all $t \geq 0$.

23. Consider the two electrical circuits with input voltages and output voltages as depicted in Fig. 3-10. Compute the transfer functions $Y_1(s)/U_1(s)$ and $Y_2(s)/U_2(s)$. Then compute the transfer function of the cascade

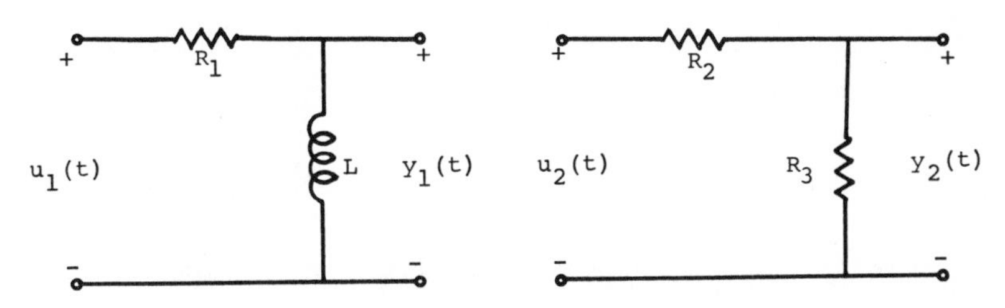

FIG. 3-10. Electrical circuits for Problem 23.

connection ($u_2(t) = y_1(t)$), $Y_2(s)/U_1(s)$ and show that the approach used in Example 13 breaks down. (This problem shows that if block diagram calculations are used, it must be verified that subsystem transfer functions are independent.)

24. Compute e^{At} for

$$A = \begin{bmatrix} \lambda & 1 & 0 \\ 0 & \lambda & 1 \\ 0 & 0 & \lambda \end{bmatrix}$$

where λ is a real number.

REMARKS AND REFERENCES

1. In considering input/output behavior, the assumption that $x_o = 0$ is made for convenience. Basically we need to assume that the initial state is the same when every input is applied. But then the response to $u(t) = 0$, $t \geq 0$ can be subtracted from the response to every input so that without loss of generality we take $x_o = 0$.

2. The usual definition of the z-transform of a discrete-time signal $f(k)$ involves viewing z as a complex variable and considering the power series $\sum_{i=0}^{\infty} f(i)z^{-i}$. If this series converges for values of z in some region, the sum, written $F(z)$, is called the z-transform of $f(k)$. Thus associated to each z-transform representation is a region of convergence for which this representation is valid. See for example the treatment in J. A. Cadzow, <u>Discrete-</u>

Time Systems, Prentice-Hall, Englewood Cliffs, N.J., 1973.

Our approach is completely algebraic. No associated conditions appear with the z-transform representation and z is viewed as only a symbol which can be manipulated according to certain rules. Although the power series definition allows the consideration of z-transforms which are not proper rational functions, this generality is not very useful in linear system theory. We have traded some mathematical generality in order to be free of convergence conditions.

3. The s-transform (Laplace Transform) of a continuous-time signal f(t) is usually defined as an integral transform with a complex variable: $F(s) = \int_0^\infty f(t)e^{-st}\,dt$. Again conditions must be placed on the value of s so as to guarantee that the integral converges for a specified f(t). By using a completely algebraic approach we have sacrificed generality by restricting ourselves to rational s-transforms but we are free of convergence conditions. (The restriction to analytic signals will be removed in Chap. 7).

4. The algebraic definitions emphasize the similarity of the z and s transforms. That the formal series definition of the z-transform begins with z^0 while the s-transform begins with s^{-1} can be explained historically. Early work on the Laplace transform used the definition $F(s) = s\int_0^\infty f(t)e^{-st}dt$ which corresponds to beginning the formal series with s^0. See for example N. W. McLachlan, Modern Operational Calculus, Macmillan and Co., New York, 1948. Each formulation has particular advantages; the modern definition is probably more convenient for advanced treatments involving generalized functions such as the impulse.

5. Another algebraic approach to transform theory is to let z^{-1} denote the unit delay operation and s^{-1} the integration operation. See W. H. Huggins and D. R. Entwisle, Introductory Systems and Design, Blaisdell, Waltham,

Mass., 1968. Another formulation can be found in D. H. Moore, *Heaviside Operational Calculus*, American Elsevier, New York, 1971. For a rigorous introduction to the mathematical concept of formal series, see I. Niven, "Formal Power Series," *American Mathematical Monthly*, vol. 6, no. 8, 1969.

6. There are several methods for systematizing the block diagram manipulations in Sec. 3.5. A popular method has been the signal flow graph representation and Mason's Rule as presented in S. J. Mason and H. J. Zimmerman, *Electronic Circuits, Signals and Systems*, J. Wiley, New York, 1960.

7. There are several advanced mathematical approaches to the rigorous treatment of the impulse function. The following two volumes present a brief and readable account of those due to Mikusinski and Schwartz, respectively. A. Erdelyi, *Operational Calculus and Generalized Functions*, Holt, Rinehart, and Winston, New York, 1962. M. J. Lighthill, *Introduction to Fourier Analysis and Generalized Functions*, Cambridge University Press, 1958. While the approximations discussed in Sec. 3.6 are often useful, it is this author's view that the usefulness of the impulse function in basic linear system theory is slight and does not warrant the effort required for a rigorous treatment.

TABLE 3-1

z - Transforms

$f(k),\ k \geq 0$	$F(z)$
$u_0(k)$	1
$u_{-1}(k)$	$\frac{z}{z - 1}$
$u_{-2}(k)$	$\frac{z}{(z - 1)^2}$
λ^k	$\frac{z}{z - \lambda}$
$k\lambda^{k-1}$	$\frac{z}{(z - \lambda)^2}$
$\sin(\omega k)$	$\frac{z \sin(\omega)}{z^2 - 2z \cos(\omega) + 1}$
$\cos(\omega k)$	$\frac{z(z - \cos(\omega))}{z^2 - 2z \cos(\omega) + 1}$
$\lambda^k \sin(\omega k)$	$\frac{z \lambda \sin(\omega)}{z^2 - 2z\lambda \cos(\omega) + \lambda^2}$
$\lambda^k \cos(\omega k)$	$\frac{z^2 - z\lambda \cos(\omega)}{z^2 - 2z\lambda \cos(\omega) + \lambda^2}$

TABLE 3-2
s - Transforms

$f(t)$, $t \geq 0$	$F(s)$
$u_{-1}(t)$	$\frac{1}{s}$
$u_{-2}(t)$	$\frac{1}{s^2}$
e^{at}	$\frac{1}{s - a}$
te^{at}	$\frac{1}{(s - a)^2}$
$\sin(\omega t)$	$\frac{\omega}{s^2 + \omega^2}$
$\cos(\omega t)$	$\frac{s}{s^2 + \omega^2}$
$e^{at} \sin(\omega t)$	$\frac{\omega}{(s - a)^2 + \omega^2}$
$e^{at} \cos(\omega t)$	$\frac{s - a}{(s - a)^2 + \omega^2}$

CHAPTER 4

COMPLETE REALIZATION

Given a state vector equation system description, we have seen that it is straightforward in principle to compute the input/output map system descriptions. Given an input/output map, the problem of inferring properties of a state vector equation description is much more complex. For example a unique state vector equation cannot be obtained since any change of variables in the state vector equation leaves the input/output map unchanged.

In this chapter we consider in detail the problem of determining and characterizing those state vector equations corresponding to a prescribed input/output map. Since the structure of the problem is identical for the discrete-time and continuous-time cases, a common framework for the problem is first established. Then we solve the problem of finding least dimension state vector equations corresponding to a given input/output map.

4.1. EXAMPLES OF COMPLETE REALIZATIONS

To indicate the situations that can arise and to motivate the significance of the complete realization problem, we will consider several examples.

Example 1 The following bucket systems have unity parameter values. The single bucket system shown in Fig. 4-1 is described by the state vector equation

$$\dot{x}(t) = -x(t) + u(t)$$

$$y(t) = x(t)$$

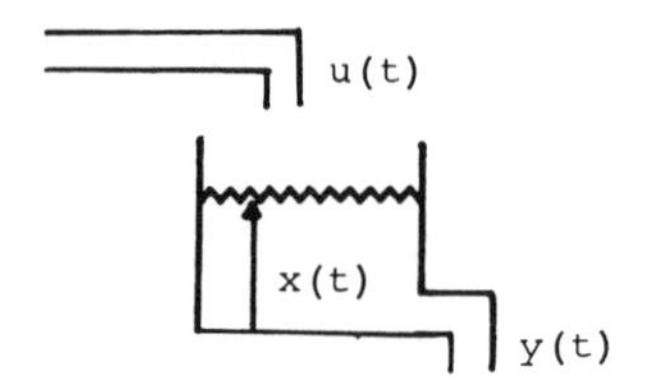

FIG. 4-1. Single bucket system.

and the transfer function

$$H(s) = \frac{1}{s + 1}$$

The three bucket system depicted in Fig. 4-2 is described by the state vector equation

$$\dot{x}(t) = \begin{bmatrix} -1 & 0 & 0 \\ 1 & -1 & 0 \\ 0 & 1 & -1 \end{bmatrix} x(t) + \begin{bmatrix} 0 \\ 1 \\ 0 \end{bmatrix} u(t)$$

$$y(t) = \begin{bmatrix} 0 & 1 & 0 \end{bmatrix} x(t)$$

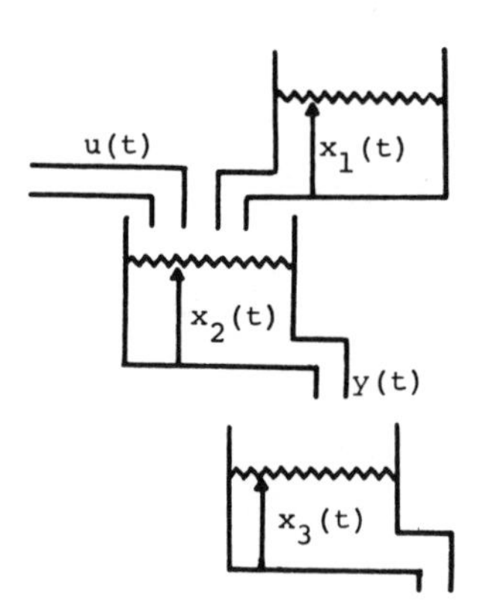

FIG. 4-2. Three bucket system.

The transfer function corresponding to this equation is

$$c(sI - A)^{-1}b + d = \frac{(s + 1)^2}{(s + 1)^3} = \frac{1}{s + 1}$$

Thus, as expected from the system diagrams, the two systems have the same input/output behavior. Of course, if the initial states are not zero the responses may be different.

Example 2 Consider the circuit shown in Fig. 4-3 where the voltage e(t) is the input and the current i(t) is the output.

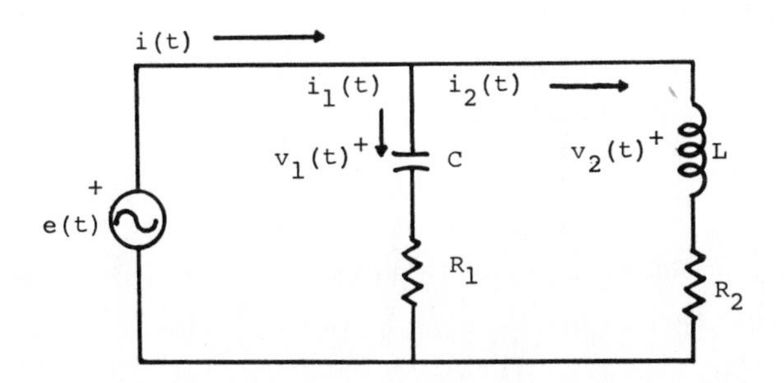

FIG. 4-3. An RLC circuit.

Choosing the state vector

$$x(t) = \begin{bmatrix} v_1(t) \\ i_2(t) \end{bmatrix}$$

gives the state vector equation

$$\dot{x}(t) = \begin{bmatrix} -1/(R_1C) & 0 \\ 0 & -R_2/L \end{bmatrix} x(t) + \begin{bmatrix} 1/(R_1C) \\ 1/L \end{bmatrix} e(t)$$

$$i(t) = \begin{bmatrix} -1/R_1 & 1 \end{bmatrix} x(t) + (1/R_1)e(t)$$

A simple calculation gives the transfer function I(s)/E(s) (the driving point admittance of the circuit):

$$\frac{I(s)}{E(s)} = \frac{(R_1^2C - L)s + R_1 - R_2}{R_1^2LCs^2 + (R_1L + R_1^2R_2C)s + R_1R_2} + \frac{1}{R_1}$$

If we pick the parameter values so that $R_1 = R_2 = R$ and $R^2C = L$, then

$$\frac{I(s)}{E(s)} = \frac{1}{R}$$

That is, with these constraints on the parameter values, the circuit has the same input/output behavior (driving point admittance) as the simple circuit shown in Fig. 4-4.

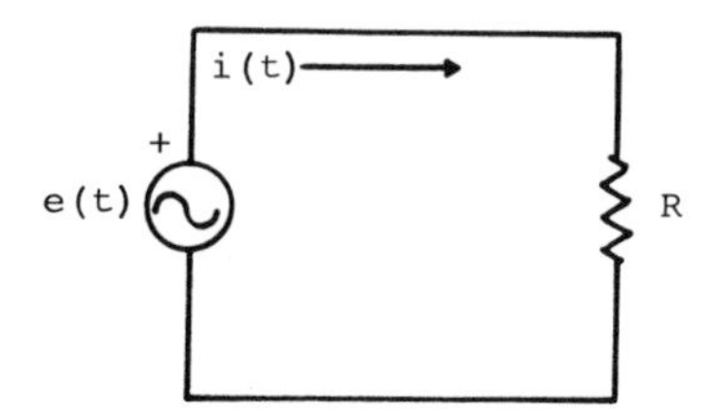

FIG. 4-4. A simple resistor circuit.

Example 3 The transfer function

$$H(z) = \frac{1}{z + 1}$$

corresponding to the state vector equation

$$x(k+1) = -x(k) + u(k)$$

$$y(k) = x(k)$$

can be written as

$$H(z) = \frac{z + 2}{(z + 1)(z + 2)} = \frac{z + 2}{z^2 + 3z + 2}$$

The corresponding RCF state vector equation can be written by inspection:

$$x(k+1) = \begin{bmatrix} 0 & 1 \\ -2 & -3 \end{bmatrix} x(k) + \begin{bmatrix} 0 \\ 1 \end{bmatrix} u(k)$$

$$y(k) = \begin{bmatrix} 2 & 1 \end{bmatrix} x(k)$$

This state vector equation has the same input/output behavior as the original.

From these examples we see that not only do state vector equations related by a nonsingular change of variable have identical input/output behavior, but state vector equations of different dimensions can have identical input/output behavior.

4.2. MINIMAL COMPLETE REALIZATIONS

We first establish a common framework for both the continuous and discrete-time cases. To do this we will always write a prescribed input/output map in special form.

In the discrete-time case, the unit pulse response corresponding to the state vector equation (A,b,c,d) is given by

$$(d,\ cb,\ cAb,\ cA^2b, \ldots) \tag{4.1}$$

The transfer function input/output map, $H(z) = c(zI - A)^{-1}b + d$ when divided out, is

$$H(z) = d + cbz^{-1} + cAbz^{-2} + cA^2bz^{-3} + \ldots \tag{4.2}$$

In the continuous-time case, the input/output map corresponding to (A,b,c,d) may be given as d and $h(t) = ce^{At}b$, $t \geq 0$. Using the series definition of the matrix exponential, h(t) can be expanded in the power series,

$$h(t) = cb + cAb\,\frac{t}{1!} + cA^2b\,\frac{t^2}{2!} + \ldots \tag{4.3}$$

As in (4.2), the transfer function $H(s) = c(sI - A)^{-1}b + d$ can be written as

$$H(s) = d + cbs^{-1} + cAbs^{-2} + cA^2bs^{-3} + \ldots \tag{4.4}$$

Thus in each situation a given input/output map specifies d and, after possible division or series expansion, an infinite sequence of scalars, $v_0 = cb$, $v_1 = cAb$, $v_2 = cA^2b$, ... The scalars v_0, v_1, ... are called the *Markov parameters* corresponding to the input/output map. We may assume without loss of generality that d = 0, since d can be determined directly from the input/output map in each case. Then the realization problem involves determining A,b, and c from the Markov parameter sequence.

<u>Definition 1</u> A state vector equation (A,b,c,0) is called a

complete realization of a prescribed input/output map if

$$cA^i b = v_i, \quad i = 0,1,... \tag{4.5}$$

It is called a *minimal complete realization* if no lower dimension complete realization exists.

Each of the examples in Sec. 4.1 contain a minimal and a nonminimal complete realization. We will concentrate on minimal complete realizations in the interest of obtaining the simplest internal system description corresponding to a prescribed external description.

The following results provide a complete characterization of minimal complete realizations of a prescribed input/output map. Note that the indeterminates z and s can be used interchangeably since the results hold for both continuous and discrete-time cases.

Theorem 1 A state vector equation (A,b,c,0) is a minimal complete realization of a prescribed input/output map $v_0, v_1, \ldots$ iff it is a complete realization and the rational function $c(zI - A)^{-1}b$ is reduced (that is, det (zI - A) and cB(z)b have no common factors, where B(z) is the classical adjoint of (zI - A)).

Proof: Suppose (A,b,c,0) is a minimal complete realization. Then $c(zI - A)^{-1}b$ is a generating function for the formal series representation,

$$cbz^{-1} + cAbz^{-2} + \ldots = v_0 z^{-1} + v_1 z^{-2} + \ldots \tag{4.6}$$

But $c(zI - A)^{-1}b$ must be the reduced generating function for, if not, then by cancelling common factors and taking, say, the RCF realization of the resulting lower degree rational function, we obtain a complete realization of lower dimension than (A,b,c,0).

Now suppose (A,b,c,0) is a complete realization and $c(zI - A)^{-1}b$ is reduced. Since the denominator polynomial of $c(zI - A)^{-1}b$ is det(zI - A) and since no cancellation can take place, the dimension of (A,b,c,0) is precisely the degree of $c(zI - A)^{-1}b$. Since $c(zI - A)^{-1}b$ is the least degree generating function of (4.6), (A,b,c,0) is a least dimension complete realization.

Theorem 2 A state vector equation is a minimal complete realization of a prescribed input/output map iff it is a CR and CO complete realization.

Proof: Suppose (A,b,c,0) is a CR and CO complete realization of dimension n. Then

$$\operatorname{rank}\begin{bmatrix} c \\ \hline cA \\ \hline \vdots \\ \hline cA^{n-1} \end{bmatrix} = \operatorname{rank}\left[b|Ab|\ldots|A^{n-1}b\right] = n$$

and since both of these are $n \times n$ matrices,

$$\operatorname{rank}\begin{bmatrix} c \\ \hline cA \\ \hline \vdots \\ \hline cA^{n-1} \end{bmatrix}\left[b|Ab|\ldots|A^{n-1}b\right]$$

$$= \operatorname{rank}\begin{bmatrix} cb & cAb & \ldots & cA^{n-1}b \\ cAb & cA^2b & \ldots & cA^nb \\ \vdots & \vdots & & \vdots \\ cA^{n-1}b & cA^nb & \ldots & cA^{2n-2}b \end{bmatrix}$$

$$= n$$

Thus the rank of the Hankel matrix corresponding to the formal series

$$cbz^{-1} + cAbz^{-2} + cA^2bz^{-3} + \ldots \qquad (4.7)$$

is at least n. However the degree of the rational function

$c(zI - A)^{-1}b$ is at most n and this rational function is a generating function for (4.7). Thus $c(zI - A)^{-1}b$ must be the least degree generating function for (4.7) and therefore must be reduced. By Theorem 1, this implies that (A,b,c,0) is a minimal complete realization.

Suppose (A,b,c,0) is a minimal complete realization of dimension n. Then by Theorem 1, $c(zI - A)^{-1}b$ is a reduced generating function of degree n of the formal series (4.7). Thus

$$n = \text{rank}\begin{bmatrix} cb & cAb & cA^2b & \cdots \\ cAb & cA^2b & cA^3b & \cdots \\ \vdots & \vdots & \vdots & \end{bmatrix}$$

$$= \text{rank}\begin{bmatrix} c \\ \hline cA \\ \hline cA^2 \\ \hline \vdots \end{bmatrix} \left[b \mid Ab \mid A^2b \mid \ldots\right]$$

Since the first matrix has n columns and the second has n rows, we have that

$$n = \text{rank}\begin{bmatrix} c \\ \hline cA \\ \hline cA^2 \\ \hline \vdots \end{bmatrix} = \text{rank}\left[b \mid Ab \mid A^2b \mid \ldots\right]$$

Using the Cayley-Hamilton Theorem as in Chap. 2, we have that for any $m > n-1$, $A^m b$ is a linear combination of b, Ab, ..., $A^{n-1}b$ and cA^m is a linear combination of c, cA, ..., cA^{n-1}. Thus

$$n = \text{rank}\begin{bmatrix} c \\ \hline cA \\ \hline \vdots \\ \hline cA^{n-1} \end{bmatrix} = \text{rank}\left[b \mid Ab \mid \ldots \mid A^{n-1}b\right]$$

which implies that (A,b,c,0) is CR and CO.

Theorem 3 Any two minimal complete realizations of a prescribed input/output map are related by a nonsingular change of variables.

Proof: Suppose the two n dimensional state vector equations, (A,b,c,d) and (F,g,h,e) are minimal complete realizations of a prescribed input/output map. Then

$$\begin{aligned} d &= e \\ cA^i b &= hF^i g, \quad i = 0,1,\ldots \end{aligned} \tag{4.8}$$

We will construct a nonsingular matrix P such that $A = P^{-1}FP$, $b = P^{-1}g$, and $c = hP$.

Since both state vector equations are complete realizations of the same input/output map, we have that

$$\begin{bmatrix} cb & cAb & \cdots & cA^{n-1}b \\ cAb & cA^2b & \cdots & cA^n b \\ \vdots & \vdots & & \vdots \\ cA^{n-1}b & cA^n b & \cdots & cA^{2n-2}b \end{bmatrix} = \begin{bmatrix} hg & hFg & \cdots & hF^{n-1}g \\ hFg & hF^2g & \cdots & hF^n g \\ \vdots & \vdots & & \vdots \\ hF^{n-1}g & hF^n g & \cdots & hF^{2n-2}g \end{bmatrix} \tag{4.9}$$

or,

$$\begin{bmatrix} c \\ \hline cA \\ \hline \vdots \\ \hline cA^{n-1} \end{bmatrix} \left[b \,|\, Ab \,|\, \ldots \,|\, A^{n-1}b \right] = \begin{bmatrix} h \\ \hline hF \\ \hline \vdots \\ \hline hF^{n-1} \end{bmatrix} \left[g \,|\, Fg \,|\, \ldots \,|\, F^{n-1}g \right] \tag{4.10}$$

Since all the n × n matrices in (4.10) are invertible by Theorem 2, we can set

$$P^{-1} = \left[b \,|\, Ab \,|\, \ldots \,|\, A^{n-1}b \right] \left[g \,|\, Fg \,|\, \ldots \,|\, F^{n-1}g \right]^{-1} \tag{4.11}$$

Then

$$P = \left[g \,|\, Fg \,|\, \ldots \,|\, F^{n-1}g \right] \left[b \,|\, Ab \,|\, \ldots \,|\, A^{n-1}b \right]^{-1} \tag{4.12}$$

Using (4.10), alternative expressions are:

$$P^{-1} = \begin{bmatrix} c \\ \hline cA \\ \hline \vdots \\ \hline cA^{n-1} \end{bmatrix}^{-1} \begin{bmatrix} h \\ \hline hF \\ \hline \vdots \\ \hline hF^{n-1} \end{bmatrix} \tag{4.13}$$

$$P = \begin{bmatrix} h \\ \hline hF \\ \hline \vdots \\ \hline hF^{n-1} \end{bmatrix}^{-1} \begin{bmatrix} c \\ \hline cA \\ \hline \vdots \\ \hline cA^{n-1} \end{bmatrix} \tag{4.14}$$

Thus we have from (4.11) that

$$\left[b|Ab|\ldots|A^{n-1}b\right] = P^{-1}\left[g|Fg|\ldots|F^{n-1}g\right] \tag{4.15}$$

the first column of which gives $b = P^{-1}g$. From (4.14) we have that

$$\begin{bmatrix} c \\ \hline cA \\ \hline \vdots \\ \hline cA^{n-1} \end{bmatrix} = \begin{bmatrix} h \\ \hline hF \\ \hline \vdots \\ \hline hF^{n-1} \end{bmatrix} P \tag{4.16}$$

the first row of which gives $c = hP$. Also, we can write that

$$\begin{bmatrix} cAb & cA^2b & \ldots & cA^nb \\ cA^2b & cA^3b & \ldots & cA^{n+1}b \\ \vdots & \vdots & & \vdots \\ cA^nb & cA^{n+1}b & \ldots & cA^{2n-1}b \end{bmatrix} = \begin{bmatrix} hFg & hF^2g & \ldots & hF^ng \\ hF^2g & hF^3g & \ldots & hF^{n+1}g \\ \vdots & \vdots & & \vdots \\ hF^ng & hF^{n+1}g & \ldots & hF^{2n-1}g \end{bmatrix} \tag{4.17}$$

which in partitioned form becomes

$$\begin{bmatrix} c \\ \hline cA \\ \hline \vdots \\ \hline cA^{n-1} \end{bmatrix} A\left[b|Ab|\ldots|A^{n-1}b\right] = \begin{bmatrix} h \\ \hline hF \\ \hline \vdots \\ \hline hF^{n-1} \end{bmatrix} F\left[g|Fg|\ldots|F^{n-1}g\right]$$

$$= \begin{bmatrix} c \\ \hline cA \\ \hline \vdots \\ \hline cA^{n-1} \end{bmatrix} P^{-1}FP\left[b|Ab|\ldots|A^{n-1}b\right]$$

Premultiplying and postmultiplying by the appropriate inverses gives

$$A = P^{-1}FP$$

which completes the proof.

This last result indicates that a minimal complete realization of a given input/output map is unique up to a change of variables. Theorem 2 provides an effective check to see if a complete realization is minimal.

A quick method for computing a minimal complete realization if the given input/output map is a transfer function $H(z)$ in factored form is the following. Obtain the reduced form of $H(z)$ by cancelling common factors from the numerator and denominator polynomials and then write either the RCF or OCF realizations as discussed in Chap. 3. They both are minimal complete realizations by Theorem 1.

PROBLEMS

1. Find complete realizations of $H(s) = 1/(s + 1)^2$ which are
   a. CR and CO
   b. CR but not CO
   c. CO but not CR
   d. neither CR nor CO

2. In Example 2 show that when $R_1 = R_2 = R$ and $R^2C = L$ the state vector equation is neither CR nor CO.

3. Show that the state vector equation

$$\dot{x}(t) = \begin{bmatrix} -1 & 0 & 1 \\ 0 & -3 & 0 \\ 0 & 0 & -2 \end{bmatrix} x(t) + \begin{bmatrix} 0 \\ 0 \\ 1 \end{bmatrix} u(t)$$

$$y(t) = \begin{bmatrix} 1 & 0 & 0 \end{bmatrix} x(t)$$

is not a minimal realization. Draw a neat SVD and explain why in terms of this diagram.

4. Show that

$$H(z) = \frac{b_{n-1}z^{n-1} + \ldots + b_1 z + b_0}{z^n + a_{n-1}z^{n-1} + \ldots + a_1 z + a_0}$$

is reduced iff

$$H(z) + d = \frac{dz^n + (b_{n-1}+da_{n-1})z^{n-1} + \ldots + (b_1+da_1)z + b_0 + da_0}{z^n + a_{n-1}z^{n-1} + \ldots + a_1 z + a_0}$$

is reduced.

5. By direct computation of the observability matrix show that the RCF realization of a degree 2 transfer function is CO iff the transfer function is reduced.

6. Consider the state vector equation (A,b,c,d) with transfer function $H(z) = c(zI - A)^{-1}b + d$. Show that each pole of H(z) is an eigenvalue of A. Show that each eigenvalue of A is a pole of H(z) iff (A,b,c,d) is CR and CO.

7. If (A,b,c,d) is a minimal realization, is it true that (A-bc,b,c-dc,d) is minimal?

8. Show that a system described by a state vector equation in distinct eigenvalue diagonal form can be decomposed into an interconnection of 4 subsystems: a CR and CO subsystem, a CR but not CO subsystem, a CO but not CR subsystem, and a subsystem that is neither CR nor CO.

REMARKS AND REFERENCES

1. We have considered what might be called the theory of "pure" complete realizations. That is, a realization is an interconnection of scalars, adders, and delayors or integrators. The problem can become considerably more difficult if a complete realization in terms of other types of physical devices is sought. While *any* proper rational function of s has a complete realization in terms of scalars, adders, and integrators, not every one, if viewed as a driving point impedance of an electrical circuit, can be realized by an RLC circuit. The use of these particular physical devices induces constraints on the set of possible driving point impedances. In fact the determination of conditions on H(s) under which it can be realized by an RLC circuit was an active area of circuit theory in the 1950's.

   To get an idea of the difficulties involved in obtaining realizations using particular physical devices, the reader might contemplate finding conditions on H(s) such that it can be realized by a bucket system.

2. A natural generalization of the problem of minimal complete realization is that of minimal partial realization. That is find a state vector equation (A,b,c,d) which corresponds to some specified portion of an input/output map. For an interesting discussion of both the complete and partial realization problems along with historical background, see R. E. Kalman, "On Minimal Partial Realizations of a Linear Input/Output Map," <u>Aspects of Network and System Theory</u>, R. E. Kalman and N. DeClaris eds., Holt, Rinehart, and Winston, New York, 1970.

3. We have concentrated on minimal realizations; those with the least number of delayors or integrators. These realizations need not be the best in other senses. For example sensitivity considerations, that is, the effect on the input/output behavior of small errors in the scalars, have not been taken into account. Certainly

some minimal complete realizations will have better sensitivity properties than others and it is possible that some nonminimal realizations will be better in this respect than some minimal realizations.

CHAPTER 5

STABILITY

Stability involves the boundedness of the response of a system to bounded inputs or to initial states. It is usually the first property that is considered when the behavior of a system is analyzed.

We first introduce a concept of internal stability for a given state vector equation description. This concept deals with the output signal due to initial states when the input signal is zero. Next a concept of input/output stability is introduced. Of course in this situation the initial state is assumed to be zero.

The important question of when these two concepts are equivalent is discussed in detail. This provides an excellent illustration of the use of material from preceding chapters. Finally we present without proof criteria for the determination of stability properties.

We present a complete treatment for the continuous-time case in this chapter. The results for the discrete-time case are analogous and thus are outlined with proofs left as exercises.

5.1. ASYMPTOTIC STABILITY

We consider the state vector equation

$$\begin{aligned} \dot{x}(t) &= Ax(t) + bu(t), \quad t \geq 0 \\ y(t) &= cx(t) + du(t), \quad x(0) = x_o \end{aligned} \tag{5.1}$$

In this section we will be only interested in the component of the output signal due to x_o.

<u>Definition 1</u> The state vector equation (5.1) is called

asymptotically stable (AS) if when $u(t) = 0$ for all $t \geq 0$, $\lim_{t\to\infty} x(t) = 0$, for every x_o.

Theorem 1 The state vector equation (5.1) is AS iff all the eigenvalues of the matrix A have negative real parts.

Proof: Since $u(t) = 0$ for all $t \geq 0$ we have that

$$x(t) = e^{At}x_o, \quad t \geq 0 \tag{5.2}$$

If the distinct eigenvalues of A are denoted by $\lambda_1, \lambda_2, \ldots, \lambda_m$ with respective multiplicities $\sigma_1, \sigma_2, \ldots, \sigma_m$, then we can write

$$e^{At} = \sum_{i=1}^{m} \sum_{j=1}^{\sigma_i} W_{ij} \frac{t^{j-1}}{(j-1)!} e^{\lambda_i t} \tag{5.3}$$

Using L'Hospital's rule and (5.3), it is shown readily that if each λ_i has a negative real part, then (5.1) is AS. Now suppose λ_i has a nonnegative real part. Then $\lim_{t\to\infty} t^{\sigma_i - 1} e^{\lambda_i t} \neq 0$ and since $W_{i\sigma_i}$ has at least one nonzero element, we can find an x_o such that $\lim_{t\to\infty} x(t) \neq 0$.

Thus the problem of ascertaining AS for the state vector equation (5.1) involves checking the root locations of the polynomial $\det(\lambda I - A)$.

5.2. UNIFORM BOUNDED INPUT BOUNDED OUTPUT STABILITY

In this section we are concerned with the output signal due to an input signal when $x_o = 0$. Thus we can work with either the state vector equation or input/output map descriptions. We will state the definitions in terms of the state vector equation for consistency, however.

Definition 2 The state vector equation (5.1) is called *uniformly bounded input bounded output stable* (UBIBOS) if the following condition holds. For $x_c = 0$ and any given $v > 0$ there is a $q > 0$ such that for all inputs satisfying

$|u(t)| \le v$ for all $t \ge 0$, the corresponding outputs satisfy $|y(t)| \le q$ for all $t \ge 0$.

There is a condition for UBIBOS associated with both input/output maps corresponding to (5.1).

<u>Theorem 2</u> The state vector equation (5.1) is UBIBOS iff $h(t) = ce^{At}b$ satisfies

$$\int_0^\infty |h(\sigma)|\, d\sigma = p < \infty \tag{5.4}$$

<u>Proof</u>: Suppose (5.4) holds and $u(t)$ is any input signal with $|u(t)| \le v$ for all $t \ge 0$. Then

$$\begin{aligned} |y(t)| &= \left|\int_0^t h(\sigma)u(t-\sigma)\, d\sigma + du(t)\right| \\ &\le \int_0^t |h(\sigma)\, u(t-\sigma)|\, d\sigma + |du(t)| \qquad (5.5) \\ &\le \int_0^t |h(\sigma)|\, d\sigma\, v + |d|v \\ &\le (p+|d|)v \text{ for all } t \ge 0 \end{aligned}$$

and we have that (5.1) is UBIBOS.

Now suppose that (5.4) does not hold, that is

$$\lim_{t\to\infty} \int_0^t |h(\sigma)|\, d\sigma = \infty \tag{5.6}$$

We will show that (5.1) is not UBIBOS by showing that for any $q > 0$ we can find an input $u(t)$ with $|u(t)| \le 1$ for all $t \ge 0$, and a time t_q such that $|y(t_q)| \ge q$. In particular for a given q pick t_q so that

$$\int_0^{t_q} |h(\sigma)|\, d\sigma > q + |d|$$

This is always possible by (5.6). Now for $0 \le \sigma \le t_q$ let

$$u(t_q-\sigma) = \text{sgn}\,[h(\sigma)] = \begin{cases} 1, & h(\sigma) > 0 \\ 0, & h(\sigma) = 0 \\ -1, & h(\sigma) < 0 \end{cases}$$

then with this input

$$h(\sigma)u(t_q-\sigma) = |h(\sigma)|$$

so that

$$y(t_q) = \int_0^{t_q} |h(\sigma)|d\sigma + du(t_q) \geq q + |d| - |d| = q$$

Theorem 3 The state vector equation (5.1) is UBIBOS iff all the roots of the denominator polynomial of the reduced form of $c(sI - A)^{-1}b$ have negative real parts. That is, iff all poles of the transfer function have negative real parts.

Proof: The reduced form of $c(sI - A)^{-1}b$ is the s-transform of h(t). Thus each root λ of the denominator polynomial yields terms of the form $t^j e^{\lambda t}$ in h(t). Clearly (5.4) can hold iff λ has negative real part.

5.3. EQUIVALENCE OF AS AND UBIBOS

The following example indicates the difficulties that arise in relating the two concepts of stability.

Example 1 The bucket system shown in Fig. 5-1 has no outflow from the second bucket and all other parameters unity.

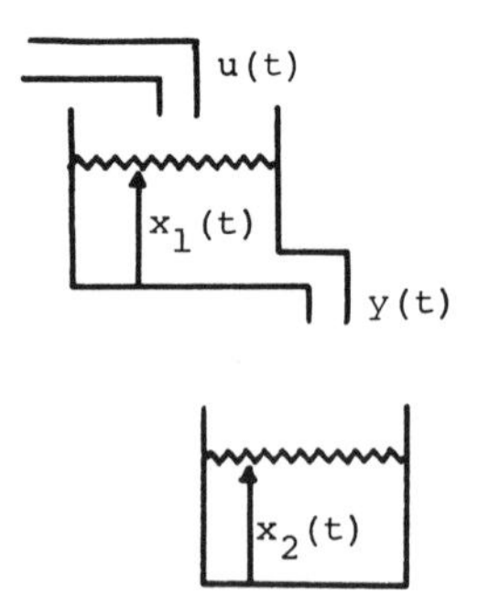

FIG. 5-1. A series bucket system

It is described by the state vector equation

$$\dot{x}(t) = \begin{bmatrix} -1 & 0 \\ 1 & 0 \end{bmatrix} x(t) + \begin{bmatrix} 1 \\ 0 \end{bmatrix} u(t)$$

$$y(t) = \begin{bmatrix} 1 & 0 \end{bmatrix} x(t)$$

Since $\det(\lambda I - A) = \lambda(\lambda + 1)$, we do not have AS. However $h(t) = e^{-t}$ so that (5.4) is satisfied and we have UBIBOS. Of course these conclusions are apparent from the system diagram in this simple case.

Example 2 A somewhat more subtle example is the RLC circuit shown in Fig. 5-2 with voltage source input $u(t)$ and output voltage $y(t)$. Choosing the inductor current $i_1(t)$ and

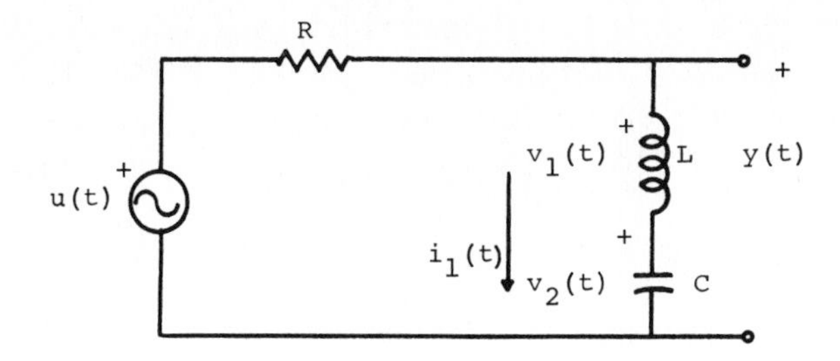

FIG. 5-2. RLC circuit

capacitor voltage $v_2(t)$ as the state variables and letting

$$x(t) = \begin{bmatrix} i_1(t) \\ v_2(t) \end{bmatrix}$$

we obtain the state vector equation description

$$\dot{x}(t) = \begin{bmatrix} -R/L & -1/L \\ 1/C & 0 \end{bmatrix} x(t) + \begin{bmatrix} 1/L \\ 0 \end{bmatrix} u(t)$$

$$y(t) = \begin{bmatrix} -R & 0 \end{bmatrix} x(t) + u(t)$$

To check AS we compute

$$\det(\lambda I - A) = \lambda^2 + \frac{R}{L}\lambda + \frac{1}{LC}$$

If R,L, and C are all positive then the quadratic formula shows that both eigenvalues have negative real parts. However if the resistance is short circuited, that is R = 0, then the circuit is not AS.

To check UBIBOS we compute

$$H(s) = c(sI - A)^{-1}b + d = \frac{s^2 + 1/LC}{s^2 + (R/L)s + 1/LC}$$

If R, L, and C are all positive this rational function is clearly reduced and we have UBIBOS. Note that if R = 0, then

$$H(s) = 1$$

and we have UBIBOS but not AS.

It is a simple task to show that AS implies UBIBOS. It is more difficult to find conditions under which the converse holds.

<u>Theorem 4</u> If the state vector equation (5.1) is AS, then it is UBIBOS.

<u>Proof</u>: From (5.3) we have that

$$h(t) = \sum_{i=1}^{m} \sum_{j=1}^{\sigma_i} cW_{ij}b \frac{t^{j-1}}{(j-1)!} e^{\lambda_i t} \tag{5.7}$$

where each λ_i is an eigenvalue of A of multiplicity σ_i. But AS implies that each λ_i has a negative real part, thus

$$\int_0^\infty |h(\sigma)|d\sigma \le \sum_{i=1}^{m} \sum_{j=1}^{\sigma_i} |cW_{ij}b| \frac{1}{(j-1)!} \int_0^\infty |\sigma^{j-1}e^{\lambda_i\sigma}| d\sigma < \infty \tag{5.8}$$

and the state vector equation is UBIBOS by Theorem 2.

<u>Theorem 5</u> If the state vector equation (5.1) is CR, CO, and UBIBOS, then it is AS.

<u>Proof</u>: If (5.1) is UBIBOS then (5.4) implies that

$$\lim_{t\to\infty} h(t) = \lim_{t\to\infty} ce^{At}b = 0$$

Also since h(t) is an analytic signal we have that

$$\lim_{t\to\infty} h^{(1)}(t) = \lim_{t\to\infty} cAe^{At}b$$

$$= \lim_{t\to\infty} ce^{At}Ab$$

$$= 0$$

$$\lim_{t\to\infty} h^{(2)}(t) = \lim_{t\to\infty} cA^2e^{At}b = \lim_{t\to\infty} cAe^{At}Ab$$

$$= \lim_{t\to\infty} ce^{At}A^2b = 0 \tag{5.9}$$

$$\vdots$$

$$\lim_{t\to\infty} h^{(2n-2)}(t) = \lim_{t\to\infty} cA^{2n-2}e^{At}b = \lim_{t\to\infty} cA^{2n-3}e^{At}Ab$$

$$= \ldots$$

$$= \lim_{t\to\infty} ce^{At}A^{2n-2}b = 0$$

Thus using partitioning we can write that

$$\lim_{t\to\infty} \begin{bmatrix} ce^{At}b & ce^{At}Ab & \ldots & ce^{At}A^{n-1}b \\ cAe^{At}b & cAe^{At}Ab & \ldots & cAe^{At}A^{n-1}b \\ \vdots & \vdots & & \vdots \\ cA^{n-1}e^{At}b & cA^{n-1}e^{At}Ab & \ldots & cA^{n-1}e^{At}A^{n-1}b \end{bmatrix}$$

$$= \lim_{t\to\infty} \begin{bmatrix} c \\ \hline cA \\ \hline \vdots \\ \hline cA^{n-1} \end{bmatrix} e^{At} \left[b|Ab|\ldots|A^{n-1}b\right] \tag{5.10}$$

$$= 0$$

Since (5.1) is CR and CO we can premultiply and postmultiply (5.10) by the appropriate inverses to obtain

$$\lim_{t\to\infty} e^{At} = 0$$

Thus (5.1) is AS and the proof is complete.

5.4. ROUTH-HURWITZ CRITERION

The actual determination of AS or UBIBOS in a particular example involves finding whether all roots of a polynomial, say,

$$p(s) = s^n + a_{n-1}s^{n-1} + \dots + a_1 s + a_0$$

have negative real parts. The following method is given without proof. Form the array with n + 1 rows shown in Table 5-1,

TABLE 5-1
Routh-Hurwitz Table for p(s).

$p(s) = s^n + a_{n-1}s^{n-1} + \dots + a_1 s + a_0$					
1	a_{n-2}	a_{n-4}	a_{n-6}	$\dots$	0
a_{n-1}	a_{n-3}	a_{n-5}	a_{n-7}	$\dots$	0
b_1	b_2	b_3	b_4	$\dots$	0
c_1	c_2	c_3	c_4	$\dots$	0
d_1	d_2	d_3	d_4	$\dots$	0
e_1	e_2	e_3	e_4	$\dots$	0
$\vdots$	$\vdots$	$\vdots$	$\vdots$		$\vdots$

where

$$b_k = \frac{-1}{a_{n-1}} \det \begin{bmatrix} 1 & a_{n-2k} \\ a_{n-1} & a_{n-2k-1} \end{bmatrix}, \quad k = 1, 2, \dots$$

$$c_k = \frac{-1}{b_1} \det \begin{bmatrix} a_{n-1} & a_{n-2k-1} \\ b_1 & b_{k+1} \end{bmatrix}, \quad k = 1, 2, \dots \tag{5.11}$$

$$d_k = \frac{-1}{c_1} \det \begin{bmatrix} b_1 & b_{k+1} \\ c_1 & c_{k+1} \end{bmatrix}, \quad k = 1, 2, \dots$$

$$e_k = \frac{-1}{d_1} \det \begin{bmatrix} c_1 & c_{k+1} \\ d_1 & d_{k+1} \end{bmatrix}, \quad k = 1, 2, \ldots$$

and so on. It is understood that the table is formed with a 0 added to the end of the first row and each row is computed out to this column. The pattern of these calculations should become clear upon working an example.

Theorem 6 (Routh-Hurwitz Criterion) All roots of the polynomial p(s) have negative real parts iff all the n+1 elements of the first column of the Routh-Hurwitz table are positive. That is, $a_{n-1} > 0$, $b_1 > 0$, $c_1 > 0, \ldots$

Example 3 The table for the polynomial

$$p(s) = s^4 + 2s^3 + 3s^2 + 4s + 5$$

is

1	3	5	0
2	4	0	0
1	5	0	0
-6	0	0	0
5	0	0	0

Since the first column has a nonpositive entry, p(s) has at least one root with a nonnegative real part.

5.5. THE DISCRETE-TIME CASE

For the state vector equation

$$\begin{aligned} x(k+1) &= Ax(k) + bu(k), \quad k = 0, 1, \ldots \\ y(k) &= cx(k) + du(k), \quad x(0) = x_o \end{aligned} \tag{5.12}$$

the following definitions are used.

Definition 3 The state vector equation (5.12) is called *asymptotically stable* (AS) if when u(k) = 0 for all k > 0, $\lim_{k\to\infty} x(k) = 0$ for every x_o.

Definition 4 The state vector equation (5.12) is called *uniformly bounded input bounded output stable* (UBIBOS) if the following condition holds. For $x_o = 0$ and any given $v > 0$ there is a $q > 0$ such that for all inputs satisfying $|u(k)| \leq v$ for all $k \geq 0$, the corresponding outputs satisfy $|y(k)| \leq q$ for all $k \geq 0$.

The proofs of the following theorems are analagous to those in the continuous-time case and thus are left as exercises.

Theorem 7 The state vector equation (5.12) is AS iff all the eigenvalues of the matrix A have magnitude less than unity.

Theorem 8 The state vector equation (5.12) is UBIBOS iff $h(k) = cA^{k-1}b$, $k \geq 1$ satisfies

$$\sum_{k=1}^{\infty} |h(k)| = p < \infty \tag{5.13}$$

Theorem 9 The state vector equation (5.12) is UBIBOS iff all the roots of the denominator polynomial of the reduced form of $c(zI - A)^{-1}b$ have magnitudes less than unity. That is, iff all poles of the transfer function have magnitudes less than unity.

Theorem 10 If the state vector equation (5.12) is AS, then it is UBIBOS. If it is CR, CO, and UBIBOS, then it is AS.

The determination of AS or UBIBOS in the discrete-time case involves finding whether all roots of a polynomial

$$p(z) = z^n + a_{n-1}z^{n-1} + \dots + a_1 z + a_0$$

have magnitude less than unity. The following test is given without proof.

Corresponding to p(z) we form Table 5-2 with 2n - 3 rows.

TABLE 5-2
Jury Table for p(z)

$p(z) = z^n + a_{n-1}z^{n-1} + \dots + a_1 z + a_0$						
a_0	a_1	a_2	$\cdots$	a_{n-2}	a_{n-1}	1
1	a_{n-1}	a_{n-2}	$\cdots$	a_2	a_1	a_0
b_0	b_1	b_2	$\cdots$	b_{n-2}	b_{n-1}	
b_{n-1}	b_{n-2}	b_{n-3}	$\cdots$	b_1	b_0	
c_0	c_1	c_2	$\cdots$	c_{n-2}		
c_{n-2}	c_{n-3}	c_{n-4}	$\cdots$	c_0		
$\vdots$	$\vdots$	$\vdots$				
r_0	r_1	r_2	r_3			
r_3	r_2	r_1	r_0			
s_0	s_1	s_2				

where

$$b_k = \det \begin{bmatrix} a_0 & a_{n-k} \\ 1 & a_k \end{bmatrix}, \quad a_n \triangleq 1,\ k = 0, 1, \dots, n-1$$

$$c_k = \det \begin{bmatrix} b_0 & b_{n-k-1} \\ b_{n-1} & b_k \end{bmatrix}, \quad k = 0, 1, \dots, n-2$$

$$d_k = \det \begin{bmatrix} c_0 & c_{n-k-2} \\ c_{n-2} & c_k \end{bmatrix}, \quad k = 0, 1, \dots, n-3$$

$$\vdots$$

$$s_k = \det \begin{bmatrix} r_0 & r_{3-k} \\ r_3 & r_k \end{bmatrix}, \quad k = 0, 1, 2$$

The pattern of these calculations should be clear after working an example.

<u>Theorem 11</u> (Jury Criterion) All roots of the polynomial

p(z) have magnitude less than unity iff the following set of n + 1 conditions is satisfied:

$$p(1) = p(z)\Big|_{z=1} > 0$$

$$(-1)^n p(-1) = (-1)^n p(z)\Big|_{z=-1} > 0$$

$$|a_0| < 1$$

$$|b_0| > |b_{n-1}|$$

$$|c_0| > |c_{n-2}|$$

$$\vdots$$

$$|s_0| > |s_2|$$

PROBLEMS

1. Devise a bucket system that is CR and UBIBOS but neither CO nor AS.
2. A continuous-time state vector equation (A,b,c,d) is called *stable* (S) if for u(t) = 0 for all $t \geq 0$, the response x(t) to any x_o is bounded. Find a necessary and sufficient condition for (A,b,c,d) to be S.
3. Show by example that S does not imply UBIBOS.
4. Check UBIBOS in Example 1 by applying Theorem 3.
5. Show that if all roots of $s^n + a_{n-1}s^{n-1} + \ldots + a_1 s + a_0$ have negative real parts, then $a_i > 0 \quad i = 0,1,\ldots,n-1$. Show by example that this condition is not sufficient.
6. Do all roots of the following polynomials have negative real parts?
   a. $s^4 + 2s^3 + 8s^2 + 4s + 3$
   b. $s^6 + 2s^5 + 3s^4 + 4s^3 + 3s^2 + 2s + 1$
7. Prove Theorem 7.
8. Prove Theorem 8.
9. Prove Theorem 9.
10. Prove Theorem 10.

11. Do all roots of the following polynomials have magnitude less than unity?

    a. $z^4 - (3/2)z^3 + 2z^2 - (1/2)z + 1/2$

    b. $z^3 + (1/2)z^2 - (1/2)z + 1/8$

12. Which of the properties AS, UBIBOS, CR, and CO does the SVD in Fig. 5-3 possess?

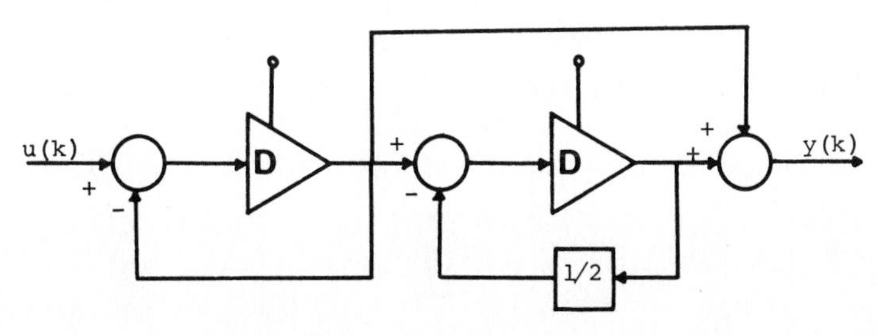

FIG. 5-3. Discrete-time SVD for Problem 12.

13. Which of the properties AS, UBIBOS, CR, and CO does the following state vector equation possess?

$$x(k+1) = \begin{bmatrix} 1/2 & 1 & 0 \\ 0 & 0 & 0 \\ 0 & 0 & -1 \end{bmatrix} x(k) + \begin{bmatrix} 0 \\ 1 \\ 0 \end{bmatrix} u(k)$$

$$y(k) = \begin{bmatrix} 1 & 0 & 0 \end{bmatrix} x(k)$$

14. Find the range of values of the parameter K for which the state vector equation

$$\dot{x}(t) = \begin{bmatrix} -2 & 1 \\ -K & -1 \end{bmatrix} x(t) + \begin{bmatrix} 0 \\ K \end{bmatrix} u(t)$$

$$y(t) = \begin{bmatrix} 1 & 0 \end{bmatrix} x(t)$$

is AS.

15. Show that for any positive integer j, $\lim_{t\to\infty} t^j e^{\lambda t} = 0$ iff λ has negative real part.

REMARKS AND REFERENCES

1. Considerable care must be exercised when reading the literature on stability. The terminology is not completely standardized and different definitions are in use. For example a continuous-time state vector equation is called BIBOS if for any input satisfying $|u(t)| \le v$ for all $t \ge 0$, there is a q such that the corresponding output satisfies $|y(t)| \le q$ for all $t \ge 0$. Note that in this definition q is allowed to depend on both v and the particular input u(t) while in Definition 2, q can depend only on v. This difference has several effects. For example if UBIBOS in Theorem 2 is replaced by BIBOS, then the necessity part of the proof is invalid.

   Perhaps surprisingly, it can be shown that BIBOS is equivalent to UBIBOS although the proof is quite difficult. We have chosen to use the UBIBOS concept because it is somewhat simpler to work with.

2. There are many useful approaches to the study of UBIBOS properties of interconnections of subsystems. The most common interconnection is a feedback system of the form shown in Fig. 5-4. Simple examples are readily constructed to show that such a system can be UBIBOS even though G(s) and/or H(s) are not UBIBOS transfer functions.

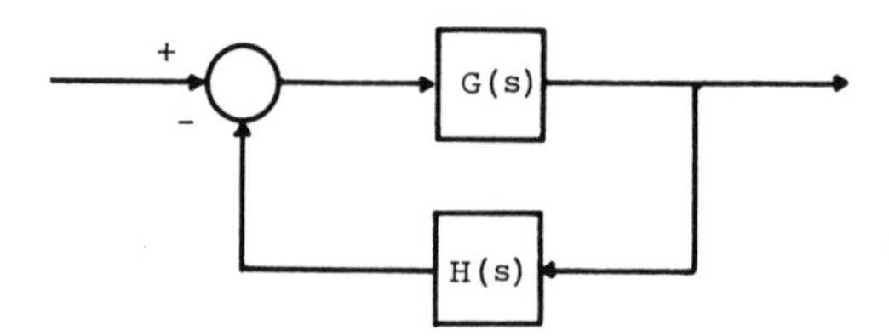

FIG. 5-4. Feedback system.

   Conversely the overall system may not be UBIBOS even though both G(s) and H(s) are UBIBOS. Since this form of system arises frequently, special techniques have been developed to analyze stability. Consult any beginning text in control systems.

3. The proofs of the Routh-Hurwitz and Jury criteria are quite complicated. See C. T. Chen, Introduction to Linear System Theory, McGraw-Hill Book Co., New York, 1970, and E. I. Jury, Theory and Applications of the z-Transform Method, J. Wiley and Sons, New York, 1964, respectively.

4. Another approach can be used to determine whether all roots of a polynomial $p(z)$ have magnitudes less than unity. Consider the bilinear transformation

$$w = \frac{z - 1}{z + 1} \qquad z = \frac{1 + w}{1 - w}$$

If z has magnitude less than unity then the corresponding w has $\mathrm{Re}\, w < 0$. Also if $|z| > 1$ then $\mathrm{Re}\, w > 0$. Thus the equation

$$p(z) \Big|_{z=\frac{1+w}{1-w}} = 0$$

has all roots with negative real parts iff $p(z)$ has all roots with magnitude less than unity. Thus we can use this transformation and the Routh-Hurwitz criterion to check stability in discrete-time systems. Clearly a reversal of this change of variable argument can be used to show that the Jury criterion can be applied to a transformed polynomial to check stability in the continuous-time case.

CHAPTER 6

COMPLETE IDENTIFICATION

In this chapter we consider the problem of determining an input/output map description for a system from measurements on the output signals corresponding to particular input signals. Of course we assume that the initial state is zero when these test input signals are applied.

There are two crucial assumptions which are made. The first is that the degree of the reduced transfer function description is known. Thus the problem of complete identification, also called parameter identification, is that of determining the coefficients in a reduced rational function of known degree. The choice of test signals, the choice of measurements, and the amount of data that must be taken are all important features of this problem.

The second assumption is that all measurements are made with perfect accuracy. Like the first assumption, this is made for simplicity in the theoretical development.

We first consider the discrete-time case. A method based on the unit pulse response which is a straightforward application of previous developments is described. Then two additional techniques are presented. Both of these yield similar types of information about the transfer function description.

The theory of complete identification in the continuous-time case is similar to that in the discrete-time case and many derivations are left as exercises.

Finally several limitations of the theory of complete identification are discussed. Simple examples are used to illustrate the complexity of the identification problem when some of the assumptions made in complete identification are relaxed.

6.1. COMPLETE IDENTIFICATION FROM THE UNIT PULSE RESPONSE

If a system is known to have a reduced transfer function description of degree n, H(z), then H(z) is readily obtainable from the unit pulse response using the results in Chap. 3. In fact only the first 2n + 1 values of the unit pulse response are needed since H(z) can be completely determined from the equations derived from

$$H(z) = \frac{b_n z^n + \dots + b_1 z + b_0}{z^n + a_{n-1} z^{n-1} + \dots + a_0} = h(0) + h(1)z^{-1} + \dots + h(2n)z^{-2n} + \dots \tag{6.1}$$

These equations are given in (3.15) and (3.16) of Chap. 3. A notable feature of this method is that stability assumptions are not needed since only a finite number of output values are used.

Some of the limitations of the complete identification problem are readily pointed out in the context of this simple method. Suppose that a discrete-time system is known to be linear (see Problem 2) but that the degree of the reduced transfer function description is unknown. Then it is impossible to determine n from input/output experiments on the system. That is, there is no indication of the number of unit pulse response values that must be measured in order to completely identify the input/output map. A simple example will clarify the nature of the difficulty.

Example 1 Suppose that a linear discrete-time system is believed to admit a transfer function description of degree 1. Suppose also that the first three values of the unit pulse response are found to be 1, 1/2, 1/4. Then we set

$$H(z) = \frac{b_1 z + b_0}{z + a_0} = 1 + \frac{1}{2} z^{-1} + \frac{1}{4} z^{-2} + \dots$$

which gives

$$H(z) = \frac{z}{z - (1/2)} \tag{6.2}$$

However if the assumption that n = 1 is incorrect, then the

system could have the description

$$H(z) = \frac{z^6 - 2z^5 + 2z - 1}{z^6 - (5/2)z^5 + z^4} = \frac{z}{z - (1/2)} + \frac{2}{z^4(z - 2)}$$

In this case we have only identified a portion of the input output map. Note that although the actual $H(z)$ is not UBIBOS, the identified $H(z)$ is UBIBOS!

Suppose we had guessed that $n = 6$. If the actual transfer function is

$$H(z) = \frac{z}{z - (1/2)} + \frac{2}{z^{13}(z - 2)}$$

then we would still obtain the same (incorrect) answer in (6.2).

No matter how large we assume n to be, it is clear that n actually could be larger. In other words no matter how many values of the unit pulse response are measured, even more might be required. We have no way of deciding when we have guessed a suitable n, or when we have sufficient data to solve the complete identification problem. Of course in a practical situation we may be able to obtain a transfer function which is suitable for certain purposes.

6.2. OTHER APPROACHES TO COMPLETE IDENTIFICATION

In this section we will discuss two additional input output experiments which can be used for complete identification. The two are similar in that they both require the assumption that $H(z)$ is UBIBOS and they both yield the value of the rational function $H(z)$ evaluated at particular numerical values of z.

Up to this point we have never regarded z as anything but a symbol to be manipulated according to certain rules. When we give z a numerical value we must be very careful that our computations are well defined. That is, we cannot evaluate z at a pole of $H(z)$.

The first experiment will be called the *exponential method*. We assume that the system can be described by a UBIBOS transfer function $H(z)$. Let λ_1 and λ_2 be real numbers with

$|\lambda_1| < 1$, $|\lambda_2| < 1$. We use the input signal defined by

$$u(k) = \lambda_1^k, \quad k \geq 0 \tag{6.3}$$

and compute

$$\sum_{k=0}^{\infty} y(k)\lambda_2^k \tag{6.4}$$

where $y(k)$ is the corresponding response. Then

$$\sum_{k=0}^{\infty} y(k)\lambda_2^k = \sum_{k=0}^{\infty} \sum_{j=0}^{k} h(k-j)\lambda_1^j \lambda_2^k \tag{6.5}$$

Since $h(k)$ is a one-sided signal, the upper limit on the inner sum can be extended to ∞ to give

$$\sum_{k=0}^{\infty} y(k)\lambda_2^k = \sum_{k=0}^{\infty} \sum_{j=0}^{\infty} h(k-j)\lambda_1^j \lambda_2^k \tag{6.6}$$

Interchanging the order of summation and then changing the summation index k to $q = k - j$ on the right side yields

$$\sum_{k=0}^{\infty} y(k)\lambda_2^k = \sum_{j=0}^{\infty} \sum_{q=-j}^{\infty} h(q)\lambda_1^j \lambda_2^{j+q} \tag{6.7}$$

Using the one-sided signal convention again and collecting terms, we have

$$\begin{aligned} \sum_{k=0}^{\infty} y(k)\lambda_2^k &= \sum_{j=0}^{\infty} (\lambda_1\lambda_2)^j \sum_{q=0}^{\infty} h(q)\lambda_2^q \\ &= \frac{1}{1 - \lambda_1\lambda_2} H(z)\Big|_{z=\frac{1}{\lambda_2}} \\ &= \frac{1}{1 - \lambda_1\lambda_2} H\left(\frac{1}{\lambda_2}\right) \end{aligned} \tag{6.8}$$

Thus from this input/output experiment we obtain the value of $H(1/\lambda_2)$.

The various series manipulations used in this development are valid since $|\lambda_1| < 1$, $|\lambda_2| < 1$, $y(k)$ is bounded, and $\sum_{k=0}^{\infty} h(k)$ converges absolutely. Note that $H(1/\lambda_2)$ is well defined as a consequence of the UBIBOS assumption and the fact that $|1/\lambda_2| > 1$.

Before discussing the problem of finding H(z) from information of this type, we will discuss the second experiment, usually called the *steady state frequency response method*. Again we assume that the system can be described by a UBIBOS transfer function H(z). It is convenient to write H(z) in the form

$$H(z) = \frac{p(z)}{q(z)} + d \tag{6.9}$$

where degree p(z) < degree q(z) and

$$q(z) = \prod_{j=1}^{m} (z - \lambda_j)^{\sigma_j} \tag{6.10}$$

with each λ_j satisfying $|\lambda_j| < 1$.

Consider the test input defined by

$$u(k) = \sin(\omega k), \quad k \geq 0 \tag{6.11}$$

By direct computation or from Table 3-1 in Chap. 3, the z-transform of this signal can be written as

$$U(z) = \frac{\sin(\omega)z}{z^2 - 2\cos(\omega)z + 1} = \frac{(1/2i)(e^{i\omega} - e^{-i\omega})z}{(z - e^{i\omega})(z - e^{-i\omega})} \tag{6.12}$$

We will show that for k very large - the so called steady state - the response corresponding to (6.11) yields the value of $H(e^{i\omega})$, that is, H(z) evaluated at $z = e^{i\omega}$.

Using the z-transform method we can write

$$Y(z) = \frac{(1/2i)(e^{i\omega} - e^{-i\omega})zp(z)}{(z - e^{i\omega})(z - e^{-i\omega})q(z)} + dU(z) \tag{6.13}$$

Expanding the first term on the right side of (6.13) in partial fractions yields

$$Y(z) = \sum_{j=1}^{m} \sum_{k=1}^{\sigma_j} \frac{K_{jk}z}{(z - \lambda_j)^k} + \frac{K_1 z}{z - e^{i\omega}} + \frac{K_2 z}{z - e^{-i\omega}} + dU(z) \tag{6.14}$$

where

$$K_1 = \left. \frac{(1/2i)(e^{i\omega} - e^{-i\omega})p(z)}{(z - e^{-i\omega})q(z)} \right|_{z=e^{i\omega}} = \frac{1}{2i} \frac{p(e^{i\omega})}{q(e^{i\omega})}$$

$$K_2 = \left. \frac{(1/2i)(e^{i\omega} - e^{-i\omega})p(z)}{(z - e^{i\omega})q(z)} \right|_{z=e^{-i\omega}} = \left(\frac{-1}{2i}\right) \frac{p(e^{-i\omega})}{q(e^{-i\omega})} \tag{6.15}$$

and the constants K_{jk} are given by similar calculations. Note that K_1 and K_2 are well defined, that is, $q(e^{\pm i\omega}) \neq 0$, since $H(z)$ is UBIBOS and $|e^{\pm i\omega}| = 1$.

Since each λ_j has magnitude less than unity, all terms in (6.14) except the last three yield components of $y(k)$ which vanish as $k \to \infty$. That is for k very large, $y(k)$ is very nearly given by

$$y_{ss}(k) = \left(\frac{1}{2i}\right) \frac{p(e^{i\omega})}{q(e^{i\omega})} e^{i\omega k} - \left(\frac{1}{2i}\right) \frac{p(e^{-i\omega})}{q(e^{-i\omega})} e^{-i\omega k} + d \sin(\omega k) \tag{6.16}$$

(The subscript "ss" is used to indicate that the steady state output, the output when k is very large, is being considered.)

To put (6.16) in better form we write

$$\frac{p(e^{i\omega})}{q(e^{i\omega})} = \mathrm{Re}\, \frac{p(e^{i\omega})}{q(e^{i\omega})} + i\, \mathrm{Im}\, \frac{p(e^{i\omega})}{q(e^{i\omega})} \tag{6.17}$$

Then, using Problem 1,

$$\frac{p(e^{-i\omega})}{q(e^{-i\omega})} = \mathrm{Re}\, \frac{p(e^{i\omega})}{q(e^{i\omega})} - i\, \mathrm{Im}\, \frac{p(e^{i\omega})}{q(e^{i\omega})} \tag{6.18}$$

Substituting (6.17) and (6.18) into (6.16) and rearranging terms, we have that $y_{ss}(k)$ is given by

$$\begin{aligned} y_{ss}(k) &= \mathrm{Re}\, \frac{p(e^{i\omega})}{q(e^{i\omega})} \frac{e^{i\omega k} - e^{-i\omega k}}{2i} + \mathrm{Im}\, \frac{p(e^{i\omega})}{q(e^{i\omega})} \frac{e^{i\omega k} + e^{-i\omega k}}{2} \\ &\quad + d \sin(\omega k) \\ &= \mathrm{Re}\big(H(e^{i\omega})\big) \sin(\omega k) + \mathrm{Im}\big(H(e^{i\omega})\big) \cos(\omega k) \\ &= |H(e^{i\omega})| \sin(\omega k + \phi), \quad \phi = \tan^{-1} \frac{\mathrm{Im}[H(e^{i\omega})]}{\mathrm{Re}[H(e^{i\omega})]} \end{aligned} \tag{6.19}$$

Thus by applying the test signal $u(k) = \sin(\omega k)$ and measuring the amplitude and phase shift of the steady state response signal, we obtain the value of $\overline{H(e^{i\omega})}$. Of course this also gives the value of $H(e^{-i\omega}) = \overline{H(e^{i\omega})}$, as shown in Problem 1.

If the system being tested is AS as well as UBIBOS then the steady state frequency response method has the advantage that it can be used without requiring that $x_o = 0$. If the initial state is nonzero, then waiting until steady state is attained also removes this component of the response.

The following theorem deals with the use of data of the type resulting from the exponential and frequency response methods to solve the complete identification problem.

Theorem 1 Suppose H(z) is a reduced UBIBOS, proper rational transfer function of degree n. Let $\lambda_1, \ldots, \lambda_{2n+1}$ be a set of distinct complex numbers with the properties that for each j, $|\lambda_j| \geq 1$, and if λ_j is complex, $\overline{\lambda_j}$ is also in the set. Then H(z) is uniquely determined by the 2n + 1 values $r_j = H(\lambda_j)$, $j = 1, \ldots, 2n+1$.

Proof: We have that H(z) must satisfy the 2n+1 equations

$$\left.\frac{b_n z^n + \ldots + b_1 z + b_0}{z^n + a_{n-1}z^{n-1} + \ldots + a_0}\right|_{z=\lambda_j} = r_j, \quad j = 1, \ldots, 2n+1 \tag{6.20}$$

Since H(z) is UBIBOS and for each j, $|\lambda_j| \geq 1$, the left side is well defined and we may write these equations as

$$b_n\lambda_j^n + \ldots + b_1\lambda_j + b_0 = r_j\lambda_j^n + r_j a_{n-1}\lambda_j^{n-1} + \ldots + r_j a_0, \; j = 1, \ldots, 2n+1 \tag{6.21}$$

or

$$b_0 + \lambda_j b_1 + \ldots + \lambda_j^n b_n - r_j a_0 - r_j\lambda_j a_1 - \ldots - r_j\lambda_j^{n-1} a_{n-1}$$

$$= r_j\lambda_j^n, \; j = 1, \ldots, 2n+1 \tag{6.22}$$

Collecting the unknown coefficients $b_0, \ldots, b_n, a_0, \ldots, a_{n-1}$ into a vector of dimension 2n + 1,

$$\begin{bmatrix} b_0 \\ \vdots \\ b_n \\ a_0 \\ \vdots \\ a_{n-1} \end{bmatrix}$$

the equations in (6.22) can be written as a single matrix equation involving this unknown vector:

$$\begin{bmatrix} 1 & \lambda_1 & \cdots & \lambda_1^n & -r_1 & -r_1\lambda_1 & \cdots & -r_1\lambda_1^{n-1} \\ 1 & \lambda_2 & \cdots & \lambda_2^n & -r_2 & -r_2\lambda_2 & \cdots & -r_2\lambda_2^{n-1} \\ \vdots & \vdots & & \vdots & \vdots & \vdots & & \vdots \\ 1 & \lambda_{2n+1} & \cdots & \lambda_{2n+1}^n & -r_{2n+1} & -r_{2n+1}\lambda_{2n+1} & \cdots & -r_{2n+1}\lambda_{2n+1}^{n-1} \end{bmatrix} \begin{bmatrix} b_0 \\ b_1 \\ \vdots \\ a_{n-1} \end{bmatrix}$$

$$= \begin{bmatrix} r_1\lambda_1^n \\ r_2\lambda_2^n \\ \vdots \\ r_{2n+1}\lambda_{2n+1}^n \end{bmatrix} \qquad (6.23)$$

Due to the assumption that $r_1, \ldots, r_{2n+1}$ correspond to the evaluations of a UBIBOS transfer function of degree n, we are assured that at least one solution for the 2n + 1 unknown coefficients in (6.23) exists. We will now show that at most one solution can exist.

Suppose that there are two solutions of (6.23), that is, two rational functions of degree n,

$$H(z) = \frac{b_n z^n + \ldots + b_1 z + b_0}{z^n + a_{n-1}z^{n-1} + \ldots + a_0}$$

$$\hat{H}(z) = \frac{\hat{b}_n z^n + \ldots + \hat{b}_1 z + \hat{b}_0}{z^n + \hat{a}_{n-1} z^{n-1} + \ldots + \hat{a}_0} \tag{6.24}$$

which satisfy the values

$$H(\lambda_j) = \hat{H}(\lambda_j) = r_j, \quad j = 1, \ldots, 2n+1$$

Then the rational function $H(z) - \hat{H}(z)$ must satisfy

$$[H(z) - \hat{H}(z)]\Big|_{z=\lambda_j} = 0, \; j = 1, \ldots, 2n+1 \tag{6.25}$$

We can write

$$H(z) - \hat{H}(z) =$$

$$\frac{(b_n z^n + \ldots + b_0)(z^n + \hat{a}_{n-1} z^{n-1} + \ldots + \hat{a}_0) - (\hat{b}_n z^n + \ldots + \hat{b}_0)(z^n + a_{n-1} z^{n-1} + \ldots + a_0)}{(z^n + \hat{a}_{n-1} z^{n-1} + \ldots + \hat{a}_0)(z^n + a_{n-1} z^{n-1} + \ldots + a_0)} \tag{6.26}$$

which makes it clear that the numerator polynomial of the rational function $H(z) - \hat{H}(z)$ is of degree at most 2n. From (6.25) we have that this polynomial must be 0 at the 2n + 1 distinct points $\lambda_1, \lambda_2, \ldots, \lambda_{2n+1}$. Thus the numerator polynomial must be the zero polynomial and we have that $H(z) - \hat{H}(z) = 0$ or $H(z) = \hat{H}(z)$. This completes the proof that (6.23) yields a unique solution.

Example 2 Suppose that the following data came from exponential tests on a system which can be described by a UBIBOS transfer function H(z) of degree 1:

$$\lambda_1 = \frac{3}{2}, \qquad r_1 = \frac{3}{2}$$

$$\lambda_2 = \frac{5}{2}, \qquad r_2 = \frac{5}{4}$$

$$\lambda_3 = \frac{7}{2}, \qquad r_3 = \frac{7}{6}$$

In this case (6.23) becomes

$$\begin{bmatrix} 1 & \frac{3}{2} & \frac{-3}{2} \\ 1 & \frac{5}{2} & \frac{-5}{4} \\ 1 & \frac{7}{2} & \frac{-7}{6} \end{bmatrix} \begin{bmatrix} b_0 \\ b_1 \\ a_0 \end{bmatrix} = \begin{bmatrix} \frac{9}{4} \\ \frac{25}{8} \\ \frac{49}{12} \end{bmatrix}$$

which has the unique solution

$$\begin{bmatrix} b_0 \\ b_1 \\ a_0 \end{bmatrix} = \begin{bmatrix} 0 \\ 1 \\ \frac{-1}{2} \end{bmatrix}$$

Thus the transfer function is

$$H(z) = \frac{b_1 z + b_0}{z + a_0} = \frac{z}{z - (1/2)}$$

Often the case when $H(z)$ is known to be strictly proper is of interest. The proof of the following result is left to Problem 6.

Corollary 1 In addition to the hypotheses of Theorem 1, suppose it is known that $H(z)$ is strictly proper $(d = 0)$. Then $H(z)$ is uniquely determined by $2n$ values, $r_j = H(\lambda_j)$, $j = 1, \ldots, 2n$.

6.3. COMPLETE IDENTIFICATION IN THE CONTINUOUS-TIME CASE

The theory of complete identification in the continuous-time case is analagous to that in the discrete-time case. However there are some differences which are significant from an application point of view. For example, if $H(s)$ is of degree n with $d = 0$ and the values of $h(0)$, $h^{(1)}(0)$, ..., $h^{(2n-1)}(0)$ are known, then solving the set of equations derived from the relation

$$\frac{b_{n-1}s^{n-1} + \ldots + b_0}{s^n + a_{n-1}s^{n-1} + \ldots + a_0}$$

$$= h(0)s^{-1} + h^{(1)}(0)s^{-2} + \ldots + h^{(2n-1)}(0)s^{-2n} + \ldots$$

gives H(s). However from the discussion in Chap. 3, Sec. 6, this approach is clearly of limited use because of the difficulty in obtaining accurately the values $h^{(j)}(0)$. This situation makes the steady state frequency response and exponential methods considerably more important.

In the exponential method we assume that H(s) is a UBIBOS transfer function. Suppose that λ_1 and λ_2 are positive real numbers and that the input $u(t) = e^{-\lambda_1 t}$ is applied with $x_o = 0$. If the response signal is y(t), then it can be shown that

$$\int_0^\infty y(t)e^{-\lambda_2 t}\,dt = \frac{1}{\lambda_1+\lambda_2}H(\lambda_2) \tag{6.27}$$

In the steady state response method, we also assume that H(s) is a UBIBOS transfer function. If the input u(t) = sin (ωt) is applied with $x_o = 0$ then the steady state response is given by

$$\begin{aligned} y_{ss}(t) &= \mathrm{Re}(H(i\omega))\sin(\omega t) + \mathrm{Im}(H(i\omega))\cos(\omega t) \\ &= |H(i\omega)|\sin(\omega t+\phi),\ \phi = \tan^{-1}\left(\frac{\mathrm{Im}[H(i\omega)]}{\mathrm{Re}[H(i\omega)]}\right) \end{aligned} \tag{6.28}$$

Again, the steady state is the response for very large values of t, after the transient terms are negligible. Also, if the system is AS, then we obtain the values of $H(i\omega)$ and $H(-i\omega) = \overline{H(i\omega)}$ regardless of the value of the initial state.

Theorem 1 and Corollary 1 of the preceding section are readily restated for the continuous-time case. It is also straightforward to see that their proofs hold with little more change than replacing z by s.

<u>Example 3</u> Suppose that for a strictly proper transfer function H(s) of degree 1, a steady state frequency response measurement gives

$$H(i3) = 4 - i3$$
$$H(-i3) = 4 + i3$$

Then, since H(s) is of the form

$$H(s) = \frac{b_0}{s + a_0}$$

we can write,

$$\left.\frac{b_0}{s + a_0}\right|_{i3} = \frac{b_0}{a_0 + i3} = 4 - i3$$

$$\left.\frac{b_0}{s + a_0}\right|_{-i3} = \frac{b_0}{a_0 - i3} = 4 + i3$$

Using Corollary 1 and setting up a 2×2 matrix equation of the form (6.23), it is found that

$$H(s) = \frac{25}{s + 4}$$

6.4. COMMENTS ON PARTIAL IDENTIFICATION

As mentioned in Sec. 1, a major impediment to the application of the theory of complete identification is that seldom if ever is n known in advance. In this case there is no indication of the amount of data that must be taken to completely identify the input/output map. A second difficulty is that we have assumed perfect measurements; the unit pulse response values or the evaluations of the transfer function are known exactly. Clearly this assumption is unrealistic.

If we allow the possibility of measurement error or the possibility that an assumed (or guessed) value for n may be too small, then the theory of identification becomes much more complicated. Three obvious situations are that the set of equations (6.23) or those derived from (6.1) may have a unique solution, several solutions, or no solution. Even in the first of these cases, the solution may be unsuitable for stability (UBIBOS) reasons. Another situation that can arise is that a solution of (6.23) may not satisfy the original data due to the occurrence of a coincident pole and zero at one of the data points. This so-called unattainable point situation is referenced in the Remarks section.

The following example and Problem 9 illustrate some of these difficulties.

Example 4 Suppose that a system which can be described by the transfer function

$$H(s) = \frac{s + 10}{(s + 1)^2}$$

is believed to be describable by a strictly proper transfer function of degree 1. Choosing the exponential method for identification, we find that $H(1) = 11/4$ and $H(2) = 4/3$. Setting

$$\left. \frac{b_0}{s + a_0} \right|_{s=1} = \frac{11}{4}$$

$$\left. \frac{b_0}{s + a_0} \right|_{s=2} = \frac{4}{3}$$

yields the matrix equation

$$\begin{bmatrix} 1 & -11/4 \\ 1 & -4/3 \end{bmatrix} \begin{bmatrix} b_0 \\ a_0 \end{bmatrix} = \begin{bmatrix} 11/4 \\ 8/3 \end{bmatrix}$$

The unique solution of this equation is

$$\begin{bmatrix} b_0 \\ a_0 \end{bmatrix} = \begin{bmatrix} 44/17 \\ -1/17 \end{bmatrix}$$

from which we have

$$H(s) = \frac{44/17}{s - (1/17)}$$

Note that this transfer function is not UBIBOS although the complete system description is UBIBOS (and the exponential method is based on this assumption).

Suppose we use different exponential tests and find that

$H(1/2) = 14/3$ and $H(2) = 4/3$. From this information we obtain the matrix equation

$$\begin{bmatrix} 1 & -14/3 \\ 1 & -4/3 \end{bmatrix} \begin{bmatrix} b_0 \\ a_0 \end{bmatrix} = \begin{bmatrix} 7/3 \\ 8/3 \end{bmatrix}$$

This equation has a unique solution and yields the transfer function

$$H(s) = \frac{56/20}{s + (1/10)}$$

Thus these measurements yield a UBIBOS transfer function description.

PROBLEMS

1. Show that if $p(z)$ is a polynomial with real coefficients and α is a complex number, then $p(\overline{\alpha}) = \overline{p(\alpha)}$. Extend this result to rational functions with real coefficients.

2. Formulate a procedure to test if an unknown system is linear. Is it possible to verify linearity with a finite number of input/output experiments?

3. For what values of ω is the signal $u(k) = \sin(\omega k)$, $k \geq 0$, a periodic signal? That is, for what values of ω does there exist a positive integer K such that $\sin(\omega k) = \sin[\omega(k+K)]$, $k \geq 0$?

4. Discuss the problems involved in measuring the phase shift in the discrete-time frequency response method.

5. Discuss the difficulties in obtaining steady state frequency response data for a bucket system.

6. Prove Corollary 1.

7. Derive the exponential method in the continuous-time case, that is, (6.27).

8. Derive the steady state frequency response method in the continuous-time case, that is, (6.28).

9. Reconsider the problem in Example 4 and suppose a steady state frequency response measurement is made at $\omega = 5$.

Find the resulting degree 1 transfer function description. Repeat this procedure with $\omega = 1$ and compare the results.

10. Suppose that from the unit pulse response values (0, 0, 1, 1/2, 1/2, ...) the transfer function description

$$H(z) = \frac{z - 1/2}{z^2(z - 1)}$$

has been determined. Find a transfer function of degree 2 that also fits this data. (This shows that a reduced transfer function description obtained from a finite amount of data is not necessarily of minimal degree.)

11. Suppose steady state frequency response tests at $\omega = 1$ and $\omega = 2$ are made on a continuous-time system which can be described by a strictly proper UBIBOS transfer function of degree 1 but that the phase shift cannot be measured. That is, with $u(t) = \sin(t)$, $y_{ss}(t) = (5/\sqrt{10})\sin(t+\phi)$, ϕ unknown, and with $u(t) = \sin(2t)$, $y_{ss}(t) = (5/\sqrt{13})$ $\sin(2t+\theta)$, θ unknown. Find out as much as you can about $H(s)$.

12. Suppose a steady state frequency response test at $\omega = 2$ is made on a continuous-time system which can be described by a strictly proper, UBIBOS transfer function of degree 1, but that only phase shift can be measured. That is, with $u(t) = \sin(2t)$, $y_{ss}(t) = K\sin(2t - 45°)$, where K is unknown. Find out as much as you can about the transfer function.

REMARKS AND REFERENCES

1. There are several types of input/output experiments which yield the values of the transfer function and its derivatives at a particular point. For example, corresponding to a discrete-time system which can be described by a UBIBOS transfer function H(z), the so called moments of the unit pulse response h(k) yield the values

$$H(1) = \sum_{k=0}^{\infty} h(k)$$

$$H^{(1)}(1) = \sum_{k=0}^{\infty} -kh(k)$$

$$\vdots$$

$$H^{(j)}(1) = (-1)^j \sum_{k=0}^{\infty} k(k+1)\dots(k+j-1)h(k),\ j = 2, 3, \dots$$

Also the exponential method can be generalized to yield transfer function derivative values at a specific point. A theorem similar to Theorem 1 can be obtained for this type of information although the proof is much more difficult. See D. R. Audley, S. L. Baumgartner, and W. J. Rugh, "Linear System Realization Based on Data Set Representations," IEEE Transactions on Automatic Control, Vol. AC-20, No. 3, 1975. Unattainable points are also mentioned.

2. Identification from steady state frequency response data in the continuous-time case is often accomplished graphically using the Bode plot technique. The Bode plot makes use of logarithmic coordinates on which straight line approximations can be used for the magnitude and phase plots of $H(i\omega)$ versus ω. This technique, which can be found in any text on control systems, is particularly convenient when only magnitude information is obtained (as in Problem 11).

3. To date relatively little is known about the structure of the partial identification problem. For the unit pulse response approach see R. E. Kalman, "On Minimal Partial Realizations of a Linear Input/Output Map," Aspects of Network and System Theory, R. E. Kalman and N. De Claris, eds., Holt, Rinehart, and Winston, New York, 1970. For other types of data, see the paper cited in Remark 1.

4. It should be clear that complete identification of the transfer function system description does not imply that

the coefficients in a particular state vector equation can be determined. If (A,b,c,d) is an n dimensional state vector equation, then there are in general $n^2 + 2n + 1$ coefficients, while in the transfer function there are $2n + 1$. The question of when the coefficients in a particular state vector equation can be determined from input/output measurements is often referred to as the *identifiability* question. Two immediate examples of state vector equations which are identifiable are RCF and OCF canonical forms. The coefficients in these particular realizations are uniquely determined by the transfer function coefficients.

CHAPTER 7

THREE SPECIAL TOPICS

Thus far we have restricted attention to those systems which admit a state vector equation description of the particular form (A,b,c,d). We will now extend this material in three different directions; each of which has important applications.

First we discuss the description of continuous-time systems which contain time delays. In this context the time delay is often called a transport lag or dead time. In the development of the description of delayed continuous-time signals we will generalize the definition of the s-transform which was given in Chap. 3.

Secondly we describe the operation of sampling a continuous-time signal. Then we develop a discrete-time description for continuous-time state vector equations with sampled input signals. Both internal and input/output descriptions are discussed.

We conclude by considering the use of a linear state vector equation as an approximation to a nonlinear state vector equation. The notions of equilibrium states and operating points are introduced and linearization is used to describe the behavior of the nonlinear equation near these points.

7.1. TIME DELAY SYSTEMS

Although delayed signals are basic in discrete-time systems, we have not considered the effect of delayed signals in the continuous-time case. Time delays occur in many different situations. For example in a bucket system the flow from, say, the input pipe at a given instant is not the flow into the first bucket until a certain time later - the time it takes for the water to fall the distance between the pipe and the bucket. We can readily imagine situations in which this effect cannot be ignored, as we have done so far.

An ideal continuous-time signal delayor with delay $T_d > 0$ is shown in Fig. 7-1 along with a typical input output pair. Suppose a one-sided signal $f(t)$ is analytic for $t \geq 0$ and has a strictly proper rational s-transform $F(s)$. With $T_d > 0$, $f(t-T_d)$ is not analytic for all $t \geq 0$ unless $f(t) = 0$ for all t. Thus we cannot apply the definition of the s-transform as given in Chap. 3. (If we went ahead, we

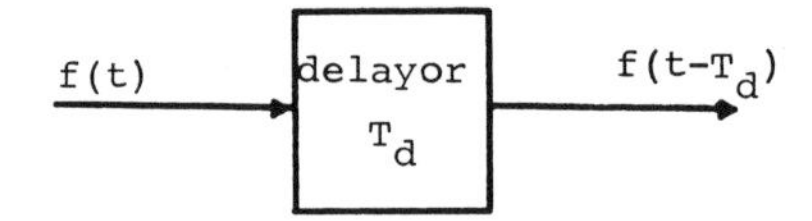

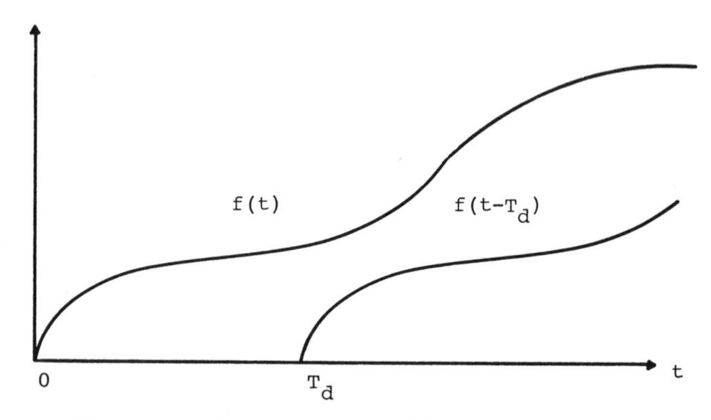

FIG. 7-1. Delayor and delayed signal.

would find that $S[f(t-T_d)] = 0$ regardless of $f(t)$ since $f(t-T_d)$ and all its derivatives are 0 at $t = 0$.)

We will generalize the definition of the s-transform as follows for signals $f(t-T_d)$ which are 0 for $t < T_d$ and analytic for $t \geq T_d$, where $T_d \geq 0$. Such a signal can be represented by its Taylor's series expansion at $t = T_d$:

$$f(t-T_d) = f(0) + f^{(1)}(0)\,\frac{t-T_d}{1!} + f^{(2)}(0)\,\frac{(t-T_d)^2}{2!} + f^{(3)}(0)\,\frac{(t-T_d)^3}{3!} + \ldots\,, \quad t \geq T_d \tag{7.1}$$

Rearranging terms we can write (7.1) in the form

$$\begin{aligned} f(t-T_d) = {} & [f(0) + f^{(1)}(0)\,\frac{t}{1!} + f^{(2)}(0)\,\frac{t^2}{2!} + f^{(3)}(0)\,\frac{t^3}{3!} + \ldots] \\ & - \frac{T_d}{1!}\,[f^{(1)}(0) + f^{(2)}(0)\,\frac{t}{1!} + f^{(3)}(0)\,\frac{t^2}{2!} + \ldots] \\ & + \frac{T_d^{\,2}}{2!}\,[f^{(2)}(0) + f^{(3)}(0)\,\frac{t}{1!} + \ldots] \\ & - \frac{T_d^{\,3}}{3!}\,[f^{(3)}(0) + \ldots] \\ & + \ldots, \quad t \geq T_d \end{aligned} \tag{7.2}$$

Substituting the place marker s^{-j} for $t^{j-1}/(j-1)!$ we obtain the formal series representation

$$\begin{aligned} f(t-T_d) = {} & [f(0)s^{-1} + f^{(1)}(0)s^{-2} + f^{(2)}(0)s^{-3} + f^{(3)}(0)s^{-4} + \ldots] \\ & - \frac{T_d}{1!}\,[f^{(1)}(0)s^{-1} + f^{(2)}(0)s^{-2} + f^{(3)}(0)s^{-3} + \ldots] \\ & + \frac{T_d^{\,2}}{2!}\,[f^{(2)}(0)s^{-1} + f^{(3)}(0)s^{-2} + \ldots] \\ & - \frac{T_d^{\,3}}{3!}\,[f^{(3)}(0)s^{-1} + \ldots] + \ldots \end{aligned} \tag{7.3}$$

It is clear that in case $T_d = 0$, this is precisely the formal series representation of $f(t)$ developed in Chap. 3. Assuming that the undelayed signal $f(t)$ has a strictly proper rational s-transform $F(s)$, we can write the generating functions for each set of terms in (7.3) as follows:

$$S[f(t-T_d)] = F(s) - \frac{T_d}{1!}[sF(s) - f(0)]$$
$$+ \frac{T_d^{\,2}}{2!}[s^2F(s) - f(0)s - f^{(1)}(0)]$$
$$- \frac{T_d^{\,3}}{3!}[s^3F(s) - f(0)s^2 - f^{(1)}(0)s - f^{(2)}(0)]$$
$$+ \cdots$$

Rearranging terms again,

$$S[f(t-T_d)] = F(s) - \frac{T_d s}{1!}F(s) + \frac{T_d^{\,2}s^2}{2!}F(s) - \frac{T_d^{\,3}s^3}{3!}F(s) + \cdots$$
$$+ \frac{T_d}{1!}f(0) - \frac{T_d^{\,2}}{2!}[f(0)s + f^{(1)}(0)] \tag{7.5}$$
$$+ \frac{T_d^{\,3}}{3!}[f(0)s^2 + f^{(1)}(0)s + f^{(2)}(0)]$$
$$+ \cdots$$

Although (7.5) contains a number of terms with nonnegative powers of s, we know from (7.3) that after division of all the terms $(T_d^j s^j/j!)F(s)$, all the nonnegative powers of s will be cancelled. That is division of all the terms in (7.5) involving F(s) will yield the set of terms

$$-\frac{T_d}{1!}f(0) + \frac{T_d^{\,2}}{2!}[f(0)s + f^{(1)}(0)]$$
$$- \frac{T_d^{\,3}}{3!}[f(0)s^2 + f^{(1)}(0)s + f^{(2)}(0)] + \cdots$$

and thus, upon simplification, only negative powers of s will remain. Because of this situation we need not carry along those terms in (7.5) which contain nonnegative powers of s. We will write

$$S[f(t-T_d)] = F(s) - \frac{T_d s}{1!}F(s) + \frac{T_d^{\,2}s^2}{2!}F(s) - \frac{T_d^{\,3}s^3}{3!}F(s) + \cdots \tag{7.6}$$

and adopt the convention that upon division in (7.6), all terms involving nonnegative powers of s are simply deleted. Clearly upon division and deletion in (7.6), we obtain precisely the formal series representation of $f(t-T_d)$ given in (7.3).

We can simplify the notation even further by writing

$$
\begin{aligned}
S[f(t-T_d)] &= [1 - \frac{T_d s}{1!} + \frac{T_d^2 s^2}{2!} - \frac{T_d^3 s^3}{3!} + \ldots]F(s) \\
&= e^{-T_d s} F(s)
\end{aligned}
\qquad (7.7)
$$

With this notation we can view the s-transform of $f(t-T_d)$ as a delay-marker $e^{-T_d s}$ multiplying the s-transform of $f(t)$. To compute the inverse s-transform of $e^{-T_d s}F(s)$ from the definition, we write it in the form (7.6) and obtain the formal series representation (7.3) by division and deletion as discussed above. But of course we need only compute the inverse s-transform of the rational function $F(s)$, $f(t)$, by the usual method and then introduce the appropriate delay to obtain $f(t-T_d)$. Again, with no delay, $T_d = 0$, this reduces naturally to the procedures discussed in Chap. 3.

It now remains to be shown that the s-transform as defined above retains properties similar to those proved in Chap. 3. This is left as an exercise for the reader. With the linearity property, the s-transform can be used to represent signals which are sums of delayed analytic signals. This extends considerably the domain of applicability of the transform representation. The inverse transform readily applies to any s-transform which can be written as a sum of terms, each of which is the product of a delay marker and a proper rational function of s.

Example 1 Using the unit ramp signal $u_{-2}(t)$ defined in Chap. 3, the signal $f(t)$ in Fig. 7-2 can be written as

$$f(t) = u_{-2}(t) - 2u_{-2}(t-1) + u_{-2}(t-2) \qquad (7.8)$$

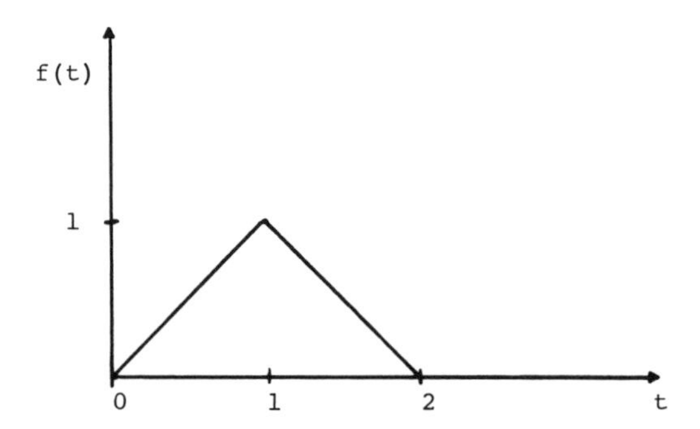

FIG. 7-2. A signal composed of unit ramps.

Thus

$$\begin{aligned} S[f(t)] &= \frac{1}{s^2} - 2e^{-s}\frac{1}{s^2} + e^{-2s}\frac{1}{s^2} \\ &= \frac{1 - 2e^{-s} + e^{-2s}}{s^2} \end{aligned} \tag{7.9}$$

One additional notational convention should be pointed out. When we write $f(t-T_d)$ or $u_{-2}(t-T_d)$ the delay interpretation is clear from the one-sided signal convention. However, when signals are written in literal form, ambiguities can appear. For example, it is not clear whether $e^{-(t-2)}$ indicates the signal e^{-t} delayed by 2 seconds or the signal e^2e^{-t} defined for all $t \geq 0$. To remove this difficulty we replace the argument t by $t-T_d$ *and* append the delayed unit step $u_{-1}(t-T_d)$ when the delay interpretation is intended. Thus we write $e^{-(t-2)}u_{-1}(t-2)$ when we describe e^{-t} delayed by 2 seconds. Similarly the notation indicates that $(t-3)^2 u_{-1}(t-3)$ is the signal t^2 delayed by 3 seconds and not the signal $(t-3)^2 = t^2 - 6t + 9$ defined for all $t \geq 0$. As a final example, note that $e^{-t}u_{-1}(t-3)$ is written in better form as $e^{-3}e^{-(t-3)}u_{-1}(t-3)$ and indicates the signal $e^{-3}e^{-t}$ delayed by 3 seconds.

We will now take the view that a time delay is a continuous-time system. It should be clear that it is linear, that is, if the responses to input signals $u_1(t)$ and $u_2(t)$

are $y_1(t)$ and $y_2(t)$ respectively, then the response to the input signal $\alpha u_1(t) + \beta u_2(t)$ is $\alpha y_1(t) + \beta y_2(t)$.

Using the s-transform description, we immediately obtain a transfer function description for a time delay of length T_d, and thus a block diagram element which is depicted in Fig. 7-3. This obeys all the usual rules for block diagram

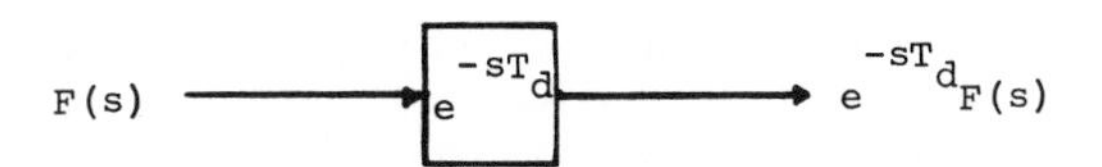

FIG. 7-3. Delayor block diagram element.

manipulation. The cascade of delays shown in Fig. 7-4

FIG. 7-4. Delayors in cascade.

yields the overall transfer function

$$\frac{Y(s)}{U(s)} = e^{-sT_1} e^{-sT_2} = e^{-s(T_1+T_2)}$$

The parallel connection shown in Fig. 7-5 yields the overall

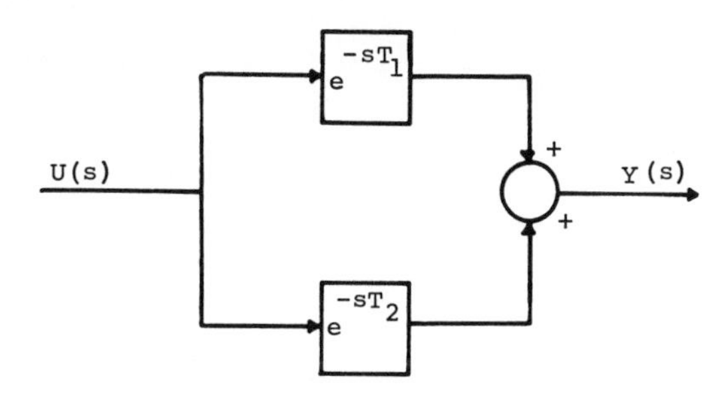

FIG. 7-5. Delayors in parallel.

transfer function

$$\frac{Y(s)}{U(s)} = e^{-sT_1} + e^{-sT_2}$$

Manipulations involving a mixture of rational transfer functions and exponential transfer functions also follow the usual block diagram rules. In particular the two block diagrams shown in Fig. 7-6 have the same overall transfer

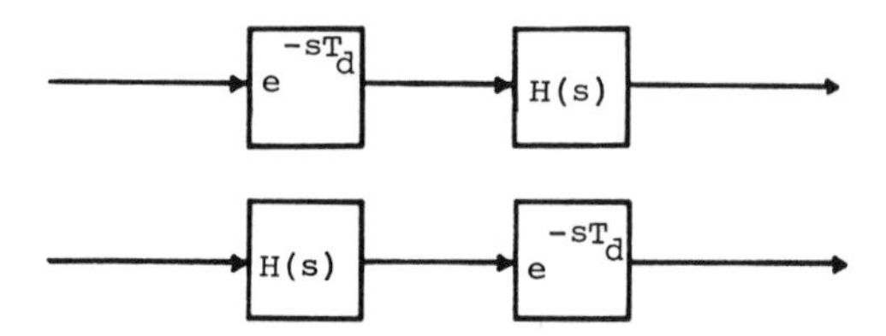

FIG. 7-6. Two block diagrams with identical transfer functions.

function. Unfortunately the situation is not as simple when a block diagram contains feedback loops, as in the diagram shown in Fig. 7-7. In this case,

$$\frac{Y(s)}{U(s)} = \frac{e^{-T_d s} H(s)}{1 + e^{-T_d s} H(s)}$$

which is not in the form of the delay marker multiplying a

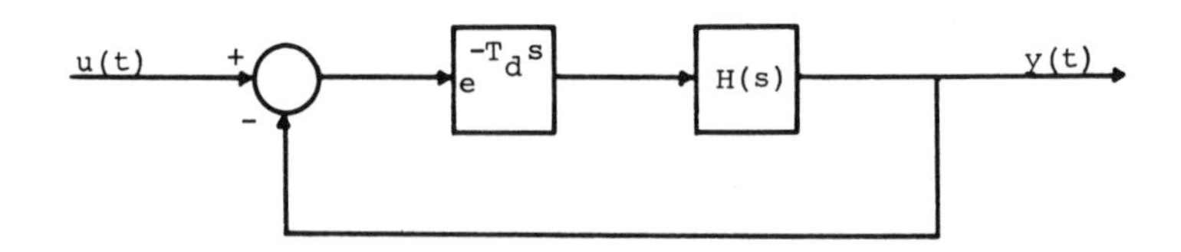

FIG. 7-7. A block diagram with feedback.

rational function. More advanced techniques must be used to analyze the input/output behavior of systems of this form.

Example 2 Suppose the signal described in Example 1 is applied to a system with transfer function

$$H(s) = \frac{1}{s+1}$$

Then, assuming zero initial state, the response signal is described by

$$\begin{aligned} Y(s) &= \frac{1 - 2e^{-s} + e^{-2s}}{s^2(s+1)} \\ &= \frac{1}{s^2(s+1)} - e^{-s}\frac{2}{s^2(s+1)} + e^{-2s}\frac{1}{s^2(s+1)} \end{aligned} \tag{7.10}$$

By partial fraction expansion it is found that the inverse s-transform of $1/s^2(s+1)$ is

$$f(t) = -1 + t + e^{-t}$$

Thus

$$y(t) = f(t) - 2f(t-1)u_{-1}(t-1) + f(t-2)u_{-1}(t-2), \quad t \geq 0 \tag{7.11}$$

The delayor is readily characterized by its response to a sinusoidal input. If $u(t) = \sin(\omega t)$, $t \geq 0$, is applied to a delayor with transfer function $e^{-T_d s}$, the response is

$$y(t) = \begin{cases} 0, & 0 \leq t < T_d \\ \sin(\omega t - \omega T_d), & t \geq T_d \end{cases} \tag{7.12}$$

Interpreting the response for $t \geq T_d$ as the steady state response, in terms of the transfer function $e^{-T_d s}$ we can write (7.12) in the form

$$\begin{aligned} y_{ss}(t) &= |e^{-i\omega T_d}| \sin(\omega t + \phi) \\ &= \sin(\omega t + \phi) \end{aligned} \tag{7.13}$$

where $\phi = \angle e^{-i\omega T_d} = -\omega T_d$. This relationship is analogous to that in the rational transfer function case. In fact, applying $u(t) = \sin(\omega t)$ to a system with transfer function $e^{-T_d s}H(s)$, where $H(s)$ is a strictly proper UBIBOS rational function, yields

$$y_{ss}(t) = |H(i\omega)| \sin(\omega t + \phi) \tag{7.14}$$

where $\phi = -\omega T_d + \tan^{-1}\left(\frac{\text{Im}[H(i\omega)]}{\text{Re}[H(i\omega)]}\right)$.

The distinguishing feature of the time delay is that the phase shift increases linearly with increasing frequency.

7.2. SAMPLED DATA SYSTEMS

The operation of sampling often occurs in modern technological systems because of the use of digital computers. The computer is a discrete-time device; it can process discrete-time signals (sequences) but cannot process or generate continuous-time signals. Sampling is a means of obtaining a discrete-time signal from a continuous-time signal. The computer is commonly used to supply input (control) signals to a continuous-time system so we will concentrate on the description of continuous-time systems with sampled inputs.

We regard the sampled version of a continuous-time signal $f(t)$ as a string of rectangular pulses occurring at the instant $t = 0, T, 2T, \ldots$, with the height of the k-th pulse given by $f(t)\big|_{t=kT} = f(kT)$. Each pulse is assumed to have width Δ, where $0 < \Delta \leq T$. We often represent a sampled signal by the sequence of its pulse heights,

$$f(kT) = (f(0), f(T), f(2T), \ldots)$$

with the sampling period T and pulse width Δ understood. The operation of sampling is represented by the system element shown in Fig. 7-8 along with a typical input/output pair. Note that sampling is a linear operation.

Suppose that the input to a continuous-time state vector equation (A,b,c,d) is a sampled signal. This form of input is too complicated to allow the use of the solution formulas derived in Chap. 1 to find $x(t)$ and $y(t)$ at each $t \geq 0$. However, we will show that a solution for $x(kT)$ and $y(kT)$, the signal values at the sampling instants, can be obtained which depends only on the input pulse heights, $u(kT)$.

The solution formulas for the state vector equation with $t = kT$ and $t = (k+1)T$ can be written as

$$x(kT) = e^{AkT}x_o + e^{AkT}\int_0^{kT} e^{-A\sigma}\, bu(\sigma)\, d\sigma$$

$$x[(k+1)T] = e^{A(k+1)T}x_o + e^{A(k+1)T}\int_0^{(k+1)T} e^{-A\sigma}\, bu(\sigma)\, d\sigma \tag{7.15}$$

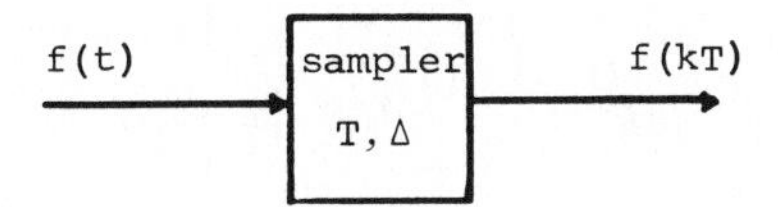

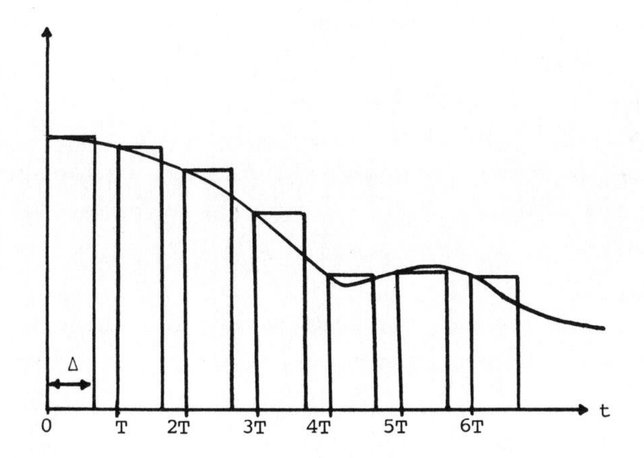

FIG. 7-8. A sampler and a sampled signal.

Multiplying both sides of the first equation in (7.15) by e^{AT} and subtracting the result from the second equation gives

$$\begin{aligned} x[(k+1)T] &= e^{AT}x(kT) + e^{A(k+1)T}\int_{kT}^{(k+1)T} e^{-A\sigma}\, bu(\sigma)\, d\sigma \\ &= e^{AT}x(kT) + \int_{kT}^{(k+1)T} e^{A[(k+1)T-\sigma]}\, bu(\sigma)\, d\sigma \end{aligned} \tag{7.16}$$

However, $u(t) = 0$, $kT+\Delta < t < (k+1)T$, and $u(t) = u(kT)$, $kT \le t \le kT+\Delta$, so that

$$x[(k+1)T] = e^{AT}x(kT) + \int_{kT}^{kT+\Delta} e^{A[(k+1)T-\sigma]}\, b\, d\sigma\, u(kT) \tag{7.17}$$

Changing the variable of integration from σ to $\tau = kT+\Delta-\sigma$

gives

$$x[(k+1)T] = e^{AT}x(kT) + \int_0^{\Delta} e^{A[T-\Delta+\tau]}\, b\, d\tau\ u(kT) \qquad (7.18)$$

$$k = 0, 1, \ldots$$

Since we can also write

$$y(kT) = cx(kT) + du(kT) \qquad (7.19)$$

we have a discrete-time state vector equation description of the behavior of the state vector and output signals at the sampling instants. It should be emphasized that this description is exact; no approximations were used to obtain (7.18) and (7.19).

Using the formula for the solution of a discrete-time state vector equation gives an iterative expression for $y(kT)$ corresponding to a given sampled signal $u(kT)$ and x_o.

Example 3 For the sampled data system and input shown in Fig. 7-9 we will determine $y(kT)$, $k = 0, 1, \ldots, 4$. Since

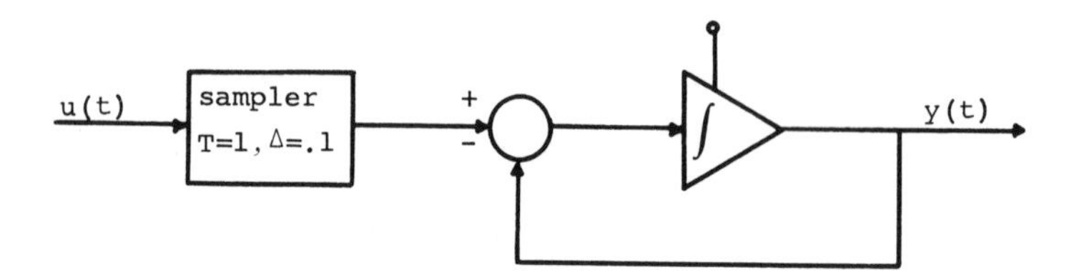

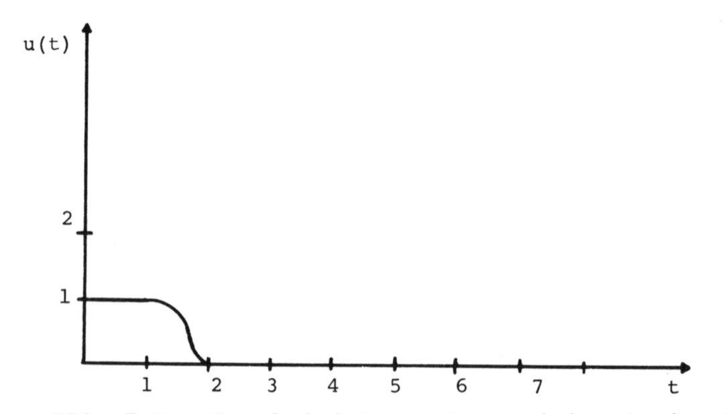

FIG. 7-9. Sampled data system and input signal.

the continuous-time state vector equation is

$$\dot{x}(t) = -x(t) + u(t)\ , \quad x(0) = 0$$

we obtain the discrete-time description

$$x[(k+1)T] = e^{-1}x(kT) + \int_0^{.1} e^{-[0.9+\tau]}\,d\tau\ u(kT)$$

$$= e^{-1}x(kT) + (e^{-0.9} - e^{-1})\ u(kT),\ x(0) = 0,\ k = 0,1,\ldots$$

Thus

$$y(kT) = y(k) = e^{-k}x_o + \sum_{j=0}^{k-1} (e^{-k+j+0.1} - e^{-k+j})u(j),\ k = 0,\ 1,\ \ldots \tag{7.21}$$

and we find that

$$\begin{aligned} y(0) &= 0 \\ y(1) &= e^{-0.9} - e^{-1} = 0.039 \\ y(2) &= e^{-1.9} - e^{-2} + e^{-0.9} - e^{-1} = 0.053 \\ y(3) &= e^{-2.9} - e^{-3} + e^{-1.9} - e^{-2} = 0.019 \\ y(4) &= e^{-3.9} - e^{-4} + e^{-2.9} - e^{-3} = 0.007 \end{aligned} \tag{7.22}$$

The z-transform can be used in conjunction with the discrete-time description of sampled data systems. In particular the input/output behavior of such systems at the sampling instants can be characterized by a z-transform transfer function. However some care must be exercised in finding the z-transform of a sampled signal $f(kT)$ since the parameter T often appears in $F(z)$. To illustrate this consider the sampled version of the unit ramp signal $u_{-2}(t)$. We represent this sampled signal by the discrete-time signal

$$u_{-2}(kT) = (0,\ T,\ 2T,\ 3T,\ \ldots)$$

Clearly the sampling period T effects the values of the sampled signal. In fact it is readily verified that

$$Z[u_{-2}(kT)] = \frac{Tz}{(z - 1)^2}$$

A table of z-transforms for common sampled signals is included at the end of the chapter. Note that with T = 1, this table reduces to that given in Chap. 3.

<u>Example 4</u> The transfer function of the discrete-time state vector equation in Example 3 is

$$H(z) = \frac{e^{-.9} - e^{-1}}{z - e^{-1}}$$

The input signal in that example is described by

$$U(z) = 1 + z^{-1} = \frac{z + 1}{z}$$

and thus

$$Y(z) = \frac{(e^{-.9} - e^{-1})(z + 1)}{z(z - e^{-1})} \tag{7.23}$$

Dividing this expression verifies (7.22)

The input/output description of sampled data systems is not always in terms of discrete-time transfer functions. It has become common to represent a sampled data system as a block diagram composed of samplers interconnected with continuous-time system transfer functions. Some care must be exercised in interpreting this representation.

The basic unit of these block diagrams is the sampled data system shown in Fig. 7-10 where G(s) is a proper rational transfer function. We can find a discrete-time transfer

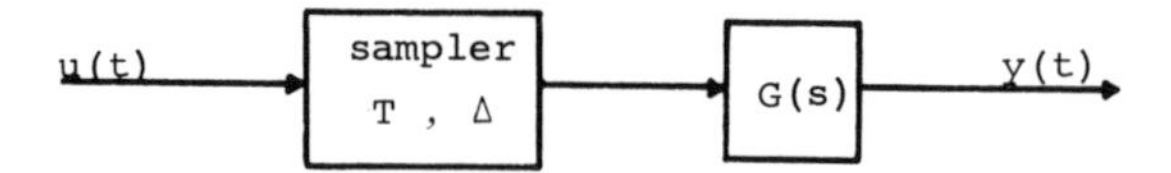

FIG. 7-10. Block diagram for sampled data system.

function description of this system as follows. Pick any minimal complete realization of G(s), for example, minimal dimension RCF. Then we can compute a discrete-time state vector equation relating u(kT) and y(kT). From this we compute a discrete-time transfer function.

Suppose a system is composed of several subsystems of this form which are interconnected through adders and scalers. It is assumed that all the samplers have identical periods T and are synchronized. Then a discrete-time transfer function can be found for each subsystem as in Fig. 7-10 and the overall transfer function can be found by ordinary discrete-time block diagram computations.

Example 5 We will obtain a discrete-time transfer function for the interconnection shown in Fig. 7-11. Picking the

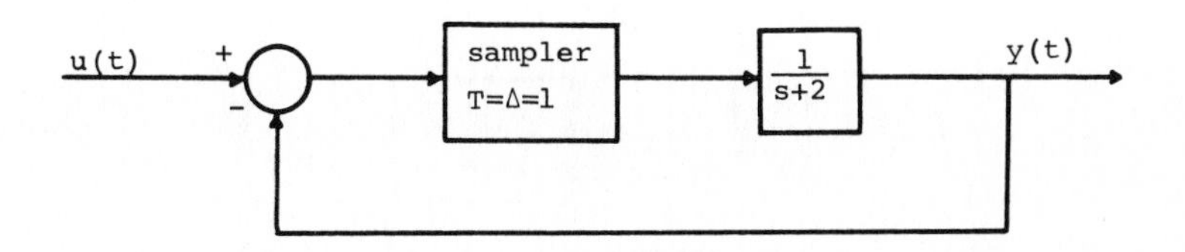

FIG. 7-11. Sampled data block diagram.

realization (A,b,c,d) = (-2,1,1,0) for 1/(s+2) we obtain the discrete-time description

$$x[(k+1)T] = e^{-2}x(kT) + \frac{1}{2}(1 - e^{-2})e(kT)$$

$$y(kT) = x(kT)$$

Since e(kT) = u(kT) - y(kT), we have the discrete time diagram depicted in Fig. 7-12. Clearly the overall discrete-

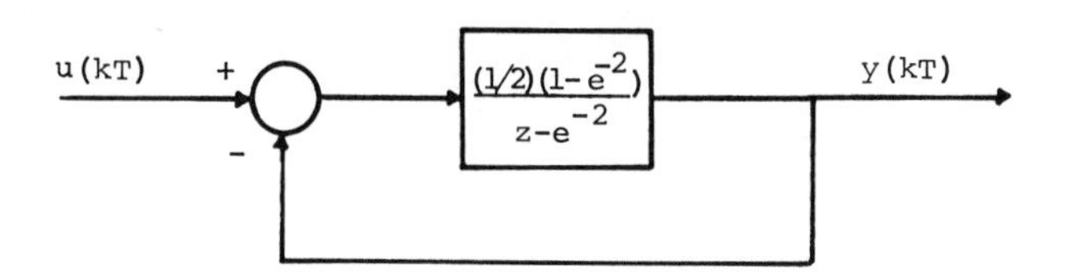

FIG. 7-12. Discrete-time block diagram.

time transfer function is

$$\frac{Y(z)}{U(z)} = \frac{(1/2)(1-e^{-2})}{z - (3/2)e^{-2} + 1/2}$$

It should be stressed that this approach is valid only when the given block diagram can be decomposed into a collection of subsystems as in Fig. 7-10 interconnected with scalars and adders. For other types of block diagrams, more advanced techniques must be used. In addition to the following example, see Remark 4.

Example 6 Consider the block diagram shown in Fig. 7-13.

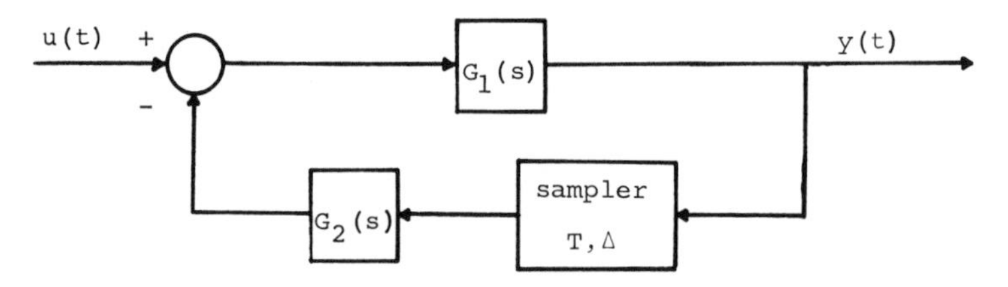

FIG. 7-13. Sampled data block diagram.

In this case there is no sampler immediately preceding $G_1(s)$ so that the approach outlined above cannot be used to find a discrete-time system description.

7.3. LINEARIZATION

In this section we will consider situations in which a linear state vector equation (A,b,c,d) can be used to approximate the behavior of a nonlinear state vector equation. We consider nonlinear equations of the general form

$$\begin{aligned} \dot{x}(t) &= f(x(t),\ u(t)), \quad t \geq 0 \\ y(t) &= g(x(t),\ u(t)), \quad x(0) = x_o \end{aligned} \tag{7.24}$$

As in the linear case, we assume x(t) is an $n \times 1$ vector signal and u(t) and y(t) are scalar signals. It should be noted that our development is readily carried over to the case of nonlinear discrete-time state vector equations of the form similar to (7.24).

The first difficulty that arises is that neither existence nor uniqueness of solutions of (7.24) is guaranteed for all $t \geq 0$.

Example 7 Consider the scalar case

$$\dot{x}(t) = 2\sqrt{x(t)} + u(t)$$

$$y(t) = x(t)$$

With $u(t) = 0$ for all $t \geq 0$ and $x(0) = 0$, there are two solutions:

$$x(t) = 0, \quad t \geq 0$$

$$x(t) = t^2, \quad t \geq 0$$

Example 8 Consider another scalar case,

$$\dot{x}(t) = x^2(t) + u(t)$$

$$y(t) = x(t)$$

With $u(t) = 0$ for all $t \geq 0$ and $x(0) = 1/2$, the solution is

$$x(t) = \frac{-1}{t - 2} \quad 0 \leq t < 2$$

Note that at $t = 2$, the solution is not defined. That is, in this case we have a solution which is not defined for all $t \geq 0$.

Rather than digress into the difficult topic of conditions for the existence and uniqueness of solutions, we will simply assume that (7.24) is such that for any given x_o and $u(t)$, there is a unique solution defined for all $t \geq 0$.

There is a special class of solutions which is of interest.

Definition 1 A pair $(\tilde{x}, \tilde{u})$ consisting of a constant $n \times 1$ vector $\tilde{x}$ and constant scalar $\tilde{u}$ is called an *operating point* of the state vector equation (7.24) if $f(\tilde{x}, \tilde{u}) = 0$. For the special case where $\tilde{u} = 0$, we call $\tilde{x}$ an *equilibrium state*.

Equilibrium states or operating points correspond to constant solutions of (7.24). That is, if $(\tilde{x}, \tilde{u})$ is an operating point and $x(0) = \tilde{x}$, $u(t) = \tilde{u}$ for all $t \geq 0$, then $x(t) = \tilde{x}$ for all $t \geq 0$. Note that this interpretation has already involved the uniqueness assumption. To see how, reconsider Example 7 and note that $\tilde{x} = 0$ is an equilibrium state.

We will also call $\tilde{y} = g(\tilde{x}, \tilde{u})$ the *operating point output*, or *equilibrium output* if $\tilde{u} = 0$.

Example 9 The series water bucket system with all parameters unity is described by

$$\dot{x}(t) = \begin{bmatrix} -1 & 0 \\ 1 & -1 \end{bmatrix} x(t) + \begin{bmatrix} 1 \\ 0 \end{bmatrix} u(t)$$

$$y(t) = \begin{bmatrix} 0 & 1 \end{bmatrix} x(t)$$

Suppose we wish to find an operating point $(\tilde{x}, \tilde{u})$ with the property that the operating point output is $\tilde{y} = 2$. Thus we must find $\tilde{u}$ and $\tilde{x}$ which satisfy

$$0 = \begin{bmatrix} -1 & 0 \\ 1 & -1 \end{bmatrix} \tilde{x} + \begin{bmatrix} 1 \\ 0 \end{bmatrix} \tilde{u}$$

$$2 = \begin{bmatrix} 0 & 1 \end{bmatrix} \tilde{x}$$

Since A is invertible in this case, for every $\tilde{u}$ there is an $\tilde{x}$ such that the first equation is satisfied, namely,

$$\tilde{x} = -\begin{bmatrix} -1 & 0 \\ 1 & -1 \end{bmatrix}^{-1} \begin{bmatrix} 1 \\ 0 \end{bmatrix} \tilde{u}$$

Thus we must find a $\tilde{u}$ which satisfies

$$2 = -\begin{bmatrix} 0 & 1 \end{bmatrix} \begin{bmatrix} -1 & 0 \\ 1 & -1 \end{bmatrix}^{-1} \begin{bmatrix} 1 \\ 0 \end{bmatrix} \tilde{u}$$

Solving these equations yields

$$\tilde{x} = \begin{bmatrix} 2 \\ 2 \end{bmatrix}, \quad \tilde{u} = 2$$

Note that the operating point input, $\tilde{u} = 2$, could be deduced from the fact that if the output is to be 2 ft^3/sec then the input has to be 2 ft^3/sec. The operating point tank levels depend upon the system parameters.

Example 10 It is readily verified that the state vector equation

$$\begin{bmatrix} \dot{x}_1(t) \\ \dot{x}_2(t) \end{bmatrix} = \begin{bmatrix} -x_1(t) + x_2(t) \\ -x_1(t) - x_2(t) + x_2^3(t) + u(t) \end{bmatrix}$$

has three equilibrium states:

$$\begin{bmatrix} 0 \\ 0 \end{bmatrix} \quad \begin{bmatrix} -\sqrt{2} \\ -\sqrt{2} \end{bmatrix} \quad \begin{bmatrix} \sqrt{2} \\ \sqrt{2} \end{bmatrix}$$

Finding equilibruim states or operating points for nonlinear state vector equations involves solving nonlinear algebraic equations. Of course this can be far from an easy task. For the linear case the situation is simpler. (See Problem 8)

Suppose that $(\tilde{x}, \tilde{u})$ is an operating point for (7.24) and we are interested in the behavior of the solutions of (7.24) for initial states x_o which are close to $\tilde{x}$ and for inputs $u(t)$ which are close to $\tilde{u}$ for all $t \geq 0$. The technique of linearization is useful in this situation under the assumption that the corresponding solutions $x(t)$ remain close to $\tilde{x}$ for all $t \geq 0$. Linearization essentially consists of replacing the functions f and g in (7.24) by the first two terms in their respective Taylor's series expansions about $\tilde{x}$ and $\tilde{u}$.

More precisely, we write the input signal of interest, $u(t)$, as

$$u(t) = \tilde{u} + u_\delta(t)$$

and the initial state of interest x_o as

$$x_o = \tilde{x} + x_{\delta o}$$

where $(\tilde{x}, \tilde{u})$ is an operating point. The corresponding solution is written as

$$x(t) = \tilde{x} + x_\delta(t)$$

Then, substituting into (7.24), we can write

$$\dot{x}(t) = \dot{x}_\delta(t) = f(\tilde{x}+x_\delta(t),\ \tilde{u}+u_\delta(t)),\ x_{\delta o} = x_o - \tilde{x}$$
$$y(t) = g(\tilde{x}+x_\delta(t),\ \tilde{u}+u_\delta(t)) \tag{7.25}$$

Assuming that the indicated derivatives exist and that the terms $u_\delta(t)$ and $x_\delta(t)$ are sufficiently small for all $t \geq 0$, we will expand the right sides of the equations in (7.25) in Taylor's series about $(\tilde{x}, \tilde{u})$ and drop all terms above first order. Although equal signs then should be replaced with approximately equal signs, we will leave this understood.

Since $f(\tilde{x}, \tilde{u}) = 0$, the i-th component of $f(x(t), u(t))$ expands as follows:

$$f_i(x(t), u(t)) = \frac{\partial f_i}{\partial x_1}(\tilde{x}, \tilde{u})x_{\delta 1}(t) + \dots + \frac{\partial f_i}{\partial x_n}(\tilde{x}, \tilde{u})x_{\delta n}(t) + \frac{\partial f_i}{\partial u}(\tilde{x}, \tilde{u})u_\delta(t)$$

Repeating this for each component and arranging the results into vector/matrix form, we have

$$\dot{x}_\delta(t) = \begin{bmatrix} \frac{\partial f_1}{\partial x_1}(\tilde{x}, \tilde{u}) & \frac{\partial f_1}{\partial x_2}(\tilde{x}, \tilde{u}) & \dots & \frac{\partial f_1}{\partial x_n}(\tilde{x}, \tilde{u}) \\ \frac{\partial f_2}{\partial x_1}(\tilde{x}, \tilde{u}) & \frac{\partial f_2}{\partial x_2}(\tilde{x}, \tilde{u}) & \dots & \frac{\partial f_2}{\partial x_n}(\tilde{x}, \tilde{u}) \\ \vdots & \vdots & & \vdots \\ \frac{\partial f_n}{\partial x_1}(\tilde{x}, \tilde{u}) & \frac{\partial f_n}{\partial x_2}(\tilde{x}, \tilde{u}) & \dots & \frac{\partial f_n}{\partial x_n}(\tilde{x}, \tilde{u}) \end{bmatrix} x_\delta(t) \tag{7.26}$$
$$+ \begin{bmatrix} \frac{\partial f_1}{\partial u}(\tilde{x}, \tilde{u}) \\ \frac{\partial f_2}{\partial u}(\tilde{x}, \tilde{u}) \\ \vdots \\ \frac{\partial f_n}{\partial u}(\tilde{x}, \tilde{u}) \end{bmatrix} u_\delta(t)\ ,\quad x_{\delta o} = x_o - \tilde{x}$$

Similarly, since $g(\tilde{x}, \tilde{u}) = \tilde{y}$, we expand the second equation in (7.25) to obtain

$$y_\delta(t) = \left[\frac{\partial g}{\partial x_1}(\tilde{x}, \tilde{u}) \ldots \frac{\partial g}{\partial x_n}(\tilde{x}, \tilde{u})\right] x_\delta(t) + \frac{\partial g}{\partial u}(\tilde{x}, \tilde{u}) u_\delta(t) \tag{7.27}$$

Thus the small deviations from the operating point values, $x_\delta(t)$, $u_\delta(t)$, and $y_\delta(t)$, are described by a linear state vector equation of the form

$$\begin{aligned} \dot{x}_\delta(t) &= Ax_\delta(t) + bu_\delta(t), \; t \geq 0 \\ y_\delta(t) &= cx_\delta(t) + du_\delta(t), \; x_\delta(0) = x_o - \tilde{x} \end{aligned} \tag{7.28}$$

When a solution to (7.28) is obtained, the nonlinear state vector equation response is approximated by

$$\begin{aligned} x(t) &= \tilde{x} + x_\delta(t) \quad , \quad t \geq 0 \\ y(t) &= \tilde{y} + y_\delta(t) \quad , \quad t \geq 0 \end{aligned} \tag{7.29}$$

Example 11 Consider the series water bucket system with all parameters unity but with the flow through each orifice given by the hyperbolic tangent of the water depth. Then the state vector equation description is

$$\begin{bmatrix} \dot{x}_1(t) \\ \dot{x}_2(t) \end{bmatrix} = \begin{bmatrix} -\tanh x_1(t) + u(t) \\ \tanh x_1(t) - \tanh x_2(t) \end{bmatrix}$$

$$y(t) = \tanh x_2(t)$$

Suppose we are interested in the behavior of this system near the operating point

$$\tilde{x} = \begin{bmatrix} 2 \\ 2 \end{bmatrix}, \qquad \tilde{u} = \tanh 2 \;, \quad \tilde{y} = \tanh 2$$

Calculating the partial derivatives indicated in (7.26) and (7.27) gives the linearized state vector equation

$$\dot{x}_\delta(t) = \begin{bmatrix} \operatorname{sech}^2 2 & 0 \\ -\operatorname{sech}^2 2 & \operatorname{sech}^2 2 \end{bmatrix} x_\delta(t) + \begin{bmatrix} 1 \\ 0 \end{bmatrix} u_\delta(t)$$

$$y_\delta(t) = [0 \quad -\text{sech}^2\ 2]x_\delta(t) \;, \quad x_{\delta o} = x_o - \begin{bmatrix} 2 \\ 2 \end{bmatrix}$$

Some comments are in order concerning the various assumptions upon which the linearized equation depends for validity. It is clearly difficult to verify a priori that x(t) remains close to $\tilde{x}$ for all $t \geq 0$. However the stability properties of the linearized equation give an indication. If the state vector equation (7.28) is AS, then one expects a reasonable approximation. However in the case where (7.28) is unstable, $x_\delta(t)$ cannot be expected to remain small. Of course another question concerns how small the small quantitites need be in order to have a reasonable approximation. A rigorous answer to this question is very difficult, so we simply will remark that in many situations the linearization technique performs quite well.

PROBLEMS

1. Find the s-transform of the signal shown in Fig. 7-14.

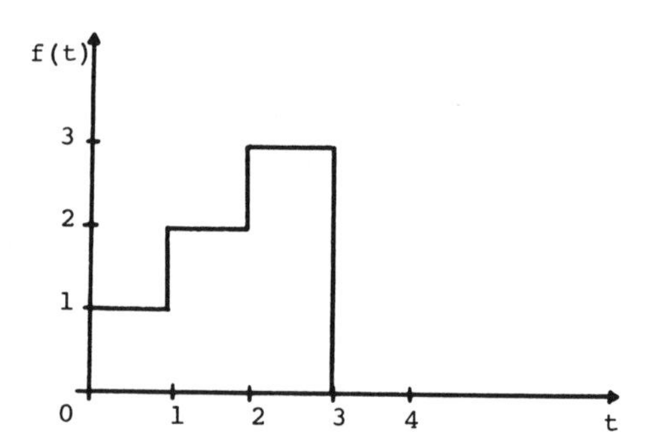

FIG. 7-14. Signal for Problem 1.

2. Find the output response of the block diagram shown in Fig. 7-15 if the input signal is that in Problem 1.

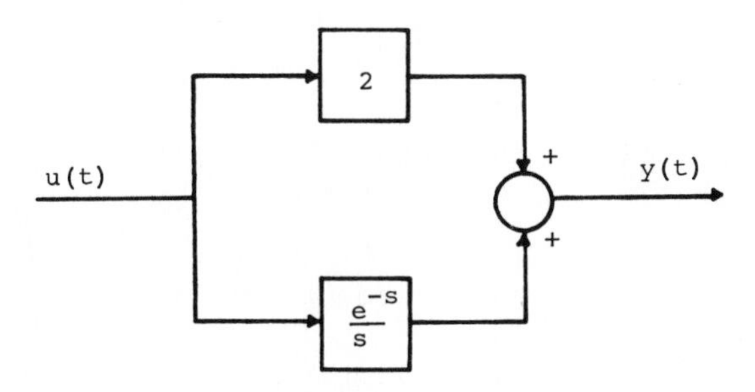

FIG. 7-15. Time delay block diagram.

3. Consider a continuous-time state vector equation in distinct eigenvalue diagonal form and suppose the input is a sampled signal. Show that the discrete-time state vector equation description is AS iff the continuous-time state vector equation is AS.

4. Find the response of the system $H(s) = 1/(s+2)$ to the input

$$u(t) = \begin{cases} e^{-t}, & 0 \le t \le 3 \\ 0, & t > 3 \end{cases}$$

5. Suppose that the input to the system in Example 3 is $u(t) = \sin(\pi t)$, $t \ge 0$, and that $x_o = 0$. Determine $y(kT)$, $k = 0,1,\ldots$.

6. Consider the state vector equation

$$\dot{x}(t) = \begin{bmatrix} \lambda_1 & 0 \\ 0 & \lambda_2 \end{bmatrix} x(t) + \begin{bmatrix} b_1 \\ b_2 \end{bmatrix} u(t)$$

where $\lambda_1 \ne \lambda_2$. Suppose $u(t)$ is a sampled signal with $T = \Delta$. Find necessary and sufficient conditions for the discrete-time description to be CR.

7. Linearize the nonlinear state vector equation in Example 10 about each equilibrium state. In each case, determine if the linearized equation is AS.

8. For the linear state vector equation $\dot{x}(t) = Ax(t) + bu(t)$, determine conditions under which there exists a unique

operating point for each $\tilde{u}$.

9. Consider the cone shaped bucket depicted in Fig. 7-16. The cone is such that when $x(t) = 1$, the surface area of the water is 4. The orifice is such that $y(t) = (1/3) x(t)$. Find an operating point with $\tilde{x} = 2$ and find a linearized state vector equation description about this operating point.

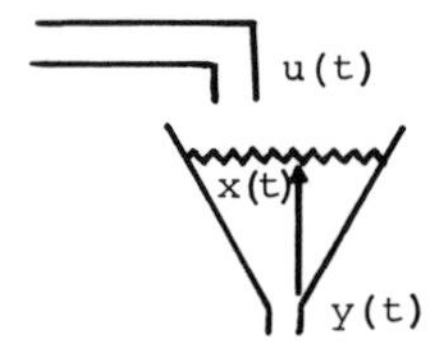

FIG. 7-16. Conical water bucket for Problem 9.

10. Find the equilibrium states for the system

$$\begin{bmatrix} \dot{x}_1(t) \\ \dot{x}_2(t) \end{bmatrix} = \begin{bmatrix} x_2(t) - 2x_1(t)x_2(t) \\ -x_1(t) + x_1^2(t) + x_2^2(t) + u(t) \end{bmatrix}$$

Linearize about each equilibrium state and comment on the stability of the linearized systems.

11. For the nonlinear discrete-time state vector equation

$$x(k+1) = f(x(k), u(k)) , \qquad x(0) = x_o$$

$$y(k) = g(x(k), u(k)) , \qquad k = 0,1,...$$

define the concepts of an operating point and an equilibrium state so that the intuitive interpretations are similar to the continuous-time case.

12. Reconsider the linear series bucket system in Example 9 and obtain a linear state vector equation description in terms of the variables

$$u_\delta(t) = u(t) - 2$$

$$x_\delta(t) = x(t) - \begin{bmatrix} 2 \\ 2 \end{bmatrix}$$

$$y_\delta(t) = y(t) - 2$$

Note that in this new description, negative values of the variables can make sense, and review Remark 3 of Chap. 2.

13. Suppose we have a continuous-time system which is a cascade connection of a delayor with $T_d = 3$ followed by a sampler with $T = \Delta = 1$ followed by the transfer function $G(s) = 1/(s+2)$. Find a discrete-time transfer function $H(z)$ which describes the behavior of this system at the sampling instants.

REMARKS AND REFERENCES

1. Using the integral definition of the s-transform (Laplace transform) it is quite easy to show that $S[f(t-T_d)] = e^{-T_d s} S[f(t)]$. In this particular case, the algebraic approach is more complicated as we have seen.

2. For small time delays, that is T_d small, a rational function approximation to $e^{-T_d s}$ is often used. The approximation is based on the frequency response properties of the delayor. The simplest approximation used is

$$e^{-i\omega T_d} \simeq \frac{1}{1 + i\omega T_d}, \quad T_d \text{ small}$$

Clearly this is rather crude since $|1/(1+i\omega T_d)|$ is not unity for all ω and the phase angle of $1/1+i\omega T_d$ does not increase linearly with increasing ω. However, over a narrow range of ω these properties are approximated as the reader can verify. A somewhat better approximation is

$$e^{-i\omega T_d} \simeq \frac{2-i\omega T_d}{2+i\omega T_d}, \quad T_d \text{ small}$$

This approximation does have unit magnitude for all ω although the phase is approximately linearly increasing only over a limited range of ω. More accurate approximations require higher degree rational functions.

3. There are many other models for the process of sampling a continuous-time signal. One of the most common is the impulse sampler model. In this model the sampled version of $f(t)$ is considered to be a string of impulse functions occurring at $t = kT$, $k = 0, 1, \ldots$ with the weight (area) of the k-th impulse given by $f(kT)$.

4. For a continuous-time system described by $G(s)$ and preceded by a sampler, there are methods for obtaining the discrete-time transfer function description directly from $G(s)$. However there are a number of subtle points involved which are often overlooked. Obtaining a state vector equation description from $G(s)$, finding the discrete-time state vector equation description, and then computing the discrete-time transfer function description is a much safer route. For the alternative method, see R. Saucedo and E. E. Schiring, _Introduction to Continuous and Digital Control Systems_, Macmillan Company, New York 1968.

5. Since a linearized state vector equation approximates the behavior of the nonlinear equation near operating points or equilibrium states, the stability properties of the linear equation imply certain local stability properties of the nonlinear equation. Also qualitative properties of the solution of the nonlinear equation near operating points or equilibrium states can be inferred, particularly in the 2-dimensional case. See Chapter 12 of K. Ogata, _Modern Control Engineering_, Prentice-Hall, Englewood Cliffs, N.J., 1970.

TABLE 7-1
Sampled Signal z-Transforms

$f(kT),\ k \geq 0$	$F(z)$
$u_0(kT)$	1
$u_{-1}(kT)$	$\dfrac{z}{z - 1}$
$u_{-2}(kT)$	$\dfrac{Tz}{(z - 1)^2}$
λ^{kT}	$\dfrac{z}{z - \lambda^T}$
$kT\lambda^{(k-1)T}$	$\dfrac{Tz}{(z - \lambda^T)^2}$
$\sin(\omega kT)$	$\dfrac{z \sin(\omega T)}{z^2 - 2z \cos(\omega T) + 1}$
$\lambda^{kT} \sin(\omega kT)$	$\dfrac{z\lambda^T \sin(\omega T)}{z^2 - 2z\lambda^T \cos(\omega T) + \lambda^{2T}}$
$\lambda^{kT} \cos(\omega kT)$	$\dfrac{z(z - \lambda^T \cos(\omega T))}{z^2 - 2z\lambda^T \cos(\omega T) + \lambda^{2T}}$

INDEX

R

S

T

U

V

W

Z